AF690594

MANAGING MOBILE TECHNOLOGIES

An Analysis from Multiple Perspectives

World Scientific Series in R&D Management

Print ISSN: 2591-7498
Online ISSN: 2591-7501

Series Editor: Tugrul U Daim *(Portland State University, USA)*

Published

Vol. 4 *Managing Mobile Technologies: An Analysis from Multiple Perspectives*
edited by Tugrul U Daim and Alexander Brem

Vol. 3 *Cooperative Innovation: Science and Technology Policy*
by Frederick Betz

Vol. 2 *Technology Roadmapping*
edited by Tugrul U Daim, Terry Oliver and Rob Phaal

Vol. 1 *Managing Technological Innovation: Tools and Methods*
edited by Tugrul U Daim

World Scientific Series in R&D Management – Vol.4

MANAGING MOBILE TECHNOLOGIES

An Analysis from Multiple Perspectives

Editors

Tugrul U Daim

Portland State University, USA

Alexander Brem

Friedrich-Alexander-University, Germany

World Scientific

NEW JERSEY • LONDON • SINGAPORE • BEIJING • SHANGHAI • HONG KONG • TAIPEI • CHENNAI • TOKYO

Published by

World Scientific Publishing Co. Pte. Ltd.
5 Toh Tuck Link, Singapore 596224
USA office: 27 Warren Street, Suite 401-402, Hackensack, NJ 07601
UK office: 57 Shelton Street, Covent Garden, London WC2H 9HE

Library of Congress Cataloging-in-Publication Data
Names: Daim, Tugrul Unsal, 1967– editor.
Title: Managing mobile technologies : an analysis from multiple perspectives / edited by Tugrul Daim (Portland State University, USA) and Alexander Brem (Friedrich-Alexander-University, Germany).
Description: New Jersey : World Scientific, [2019] | Series: World Scientific series in R&D management ; volume 4 | Includes bibliographical references.
Identifiers: LCCN 2018055245 | ISBN 9789813278158 (hc : alk. paper)
Subjects: LCSH: Mobile communication systems. | Mobile computing.
Classification: LCC TK5103.2 .M35 2019 | DDC 621.3845/6--dc23
LC record available at https://lccn.loc.gov/2018055245

British Library Cataloguing-in-Publication Data
A catalogue record for this book is available from the British Library.

Copyright © 2019 by World Scientific Publishing Co. Pte. Ltd.

All rights reserved. This book, or parts thereof, may not be reproduced in any form or by any means, electronic or mechanical, including photocopying, recording or any information storage and retrieval system now known or to be invented, without written permission from the publisher.

For photocopying of material in this volume, please pay a copying fee through the Copyright Clearance Center, Inc., 222 Rosewood Drive, Danvers, MA 01923, USA. In this case permission to photocopy is not required from the publisher.

For any available supplementary material, please visit
https://www.worldscientific.com/worldscibooks/10.1142/11208#t=suppl

Desk Editors: Anthony Alexander/Amanda Yun

Typeset by Stallion Press
Email: enquiries@stallionpress.com

To My Family, Yonca, Tolga and Eda

About the Editors

Tugrul U. Daim is a Professor and the Director of the Technology Management Doctoral Program in the Maseeh College of Engineering and Computer Science at Portland State University. He is also the Director of the Research Group on Infrastructure and Technology Management. He is a Faculty Fellow at the Institute for Sustainable Solutions. The US Department of Energy, National Science Foundation, National Cooperative Highway Research Program, and many other regional, national and international organizations have funded his research. He has published over 200 refereed papers in journals and conference proceedings. He edited more than 20 special issues in journals. He edited more than 20 books and conference proceedings. He was the adviser for 11 PhD graduates who are now in leading positions in government, industry and academia.

He is the Editor-in-Chief of *IEEE Transactions on Engineering Management. He has been at various editorial roles in journals including International Journal of Innovation and Technology Management, Technological Forecasting and Social Change, Technology in Society, Foresight, Journal of Knowledge Economy and International Journal of Innovation and Entrepreneurship*.

Prior to joining PSU, he had worked at Intel Corporation for over a decade in varying management roles. At Intel, he managed product and technology development. During his tenure at

PSU, he has acted as a Strategic Consultant to the Chief Technology Innovation Officer of Bonneville Power Administration a part of the US Department of Energy. He helped develop regional and national technology roadmaps in the energy sector. He is a member of the R&D Advisory Board for TUPRAS, the largest industrial firm in Turkey. He was also the consultant for many other international, national and regional organizations including Elsevier, Biotronik, NEEA, Energy Trust of Oregon, EPRI, ETRI, Koc Holding, Arcelik, Tofas, Ford Otosan, Kirlangic, Siemens, Mark and Spencer, and Castrol.

He is also a Visiting Professor with the Northern Institute of Technology at the Technical University of Hamburg, Harburg. He has given keynotes and distinguished speaker lectures at conferences, companies, universities and research centers around the world including Iamot, Euromot, Samsung, Helmut Schmidt University, Kuhne Logistics University, Seoul National University, Bogazici University, Koc University, University of Gaziantep, Izmir Institute of Technology, University of Pretoria, Tampere University of Technology, STEPI, EPIC at UNCC, etc.

He was given the Research Publication Award by the International Association of Management of Technology (IAMOT) and Fellow Award by the Portland International Center for Management of Engineering and Technology (PICMET) both in 2014.

Dr. Daim was the President of Omega Rho, International Honor Society in Operations Research and Management Science between the years of 2014 and 2016.

He received his BS in Mechanical Engineering from Bogazici University in Turkey, MS in Mechanical Engineering from Lehigh University in Pennsylvania, MS in Engineering Management from Portland State University, and PhD in Systems Science: Engineering Management from Portland State University in Portland Oregon.

Alexander Brem holds the Chair of Technology Management at the Friedrich-Alexander-Universität Erlangen-Nürnberg (FAU), Germany, which is located at the Nuremberg Campus of Technology (NCT). He is a CCSR International Research Associate at DeMontfort University (UK) and a Visiting Professor at the EADA Business School in Barcelona (Spain) and HHL Graduate School of Management (Germany). Moreover, he serves as an Academic Committee Member of the Center of Technological Innovation, Tsinghua University, Beijing (China). In addition, he was appointed as Honorary Professor at the University of Southern Denmark (SDU) in May 2017. Further information can be found here: https://www.tm.rw.fau.eu.

List of Contributors

Brenda Mak, College of Business, San Francisco State University, San Francisco, CA, USA, bmak@sfsu.edu

Chia-Liang Hung, Department of Information Management School of Management, National Chi Nan University, Taiwan, Republic of China, clhung@ncnu.edu.tw

Dharmesh Gupta, Shailesh J. Mehta School of Management, IIT Bombay, Mumbai, India, dharmesh@iitb.ac.in

Fumiyo N. Kondo, Department of Social Systems and Management, Graduate School of Systems and Information Engineering, The University of Tsukuba, Ibaraki, Japan, kondo@sk.tsukuba.ac.jp

Henri Isaac, University of Paris Dauphine, France, henri.isaac@dauphine.psl.edu

Jerome Chih-Lung Chou, Department of Information Management, Hwa-Hsia Institute of Technology, Taiwan, Republic of China, jerome.chou@gmail.com

Juha Winter, Aalto University, School of Science, Aalto, Finland, juha.winter@aalto.fi

Karuna Jain, National Institute of Industrial Engineering, Mumbai, India, kjain@som.iitb.ac.in

Kirankumar S. Momaya, Shailesh J. Mehta School of Management, IIT Bombay, Mumbai, India, momaya@som.iitb.ac.in

Li-Pang Ho, Department of Business Administration, National Chin-Yi University of Technology, Taiwan, Republic of China.

Miriam Martens, Detecon International GmbH, Cologne, Germany, Miriam.Martens@detecon.com

Nicole Bohn, Disability Programs and Resource Center, San Francisco State University, San Francisco, USA, nbohn@sfsu.edu

Numtip Trakulmaykee, Information and Communication Technology, Faculty of Science, Prince of Songkla University, Songkhla, Thailand, maykee_n@yahoo.com

Oliver Roll, HS Osnabrück, Faculty of Business Administration and Social Sciences, Osnabrück, Germany, roll@wi.hs-osna-brueck.de

Orawit Thinnukool, Department of Modern Management and Information Technology, College of Arts, Media and Technology, Chiang Mai University, Chiang Mai, Thailand, orawit.t@cmu.ac.th

Parichard Benrit, Department of Social Science, Faculty of Humanities and Social Sciences, Prince of Songkla University, Songkhla, Thailand, pair.ann@hotmail.com

Pattaraporn Khuwuthyakorn, Department of Modern Management and Information Technology, College of Arts, Media and Technology, Chiang Mai University, Chiang Mai, Thailand, pattaraporn.khuwuth@cmu.ac.th

Paul Beckman, College of Business, San Francisco State University, San Francisco, CA, USA, pbeckman@sfsu.edu

Qazi Mahdia Ghyas, Department of Social Systems and Management, Graduate School of Systems and Information Engineering, The University of Tsukuba, Ibaraki, Japan, s1030160@u.tsukuba.ac.jp

Robert C. Nickerson, College of Business, San Francisco State University, CA, USA, rnick@sfsu.edu

Roger Elliott, University of Fort Hare, Department of Business Management, East London, South Africa, Relliott@ufh.ac.za

Ruchita Gupta, National Institute of Industrial Engineering, Mumbai, India, ruchitagupta@nitie.edu

Ruthaychonnee Sittichai, Department of Library and Information Science, Faculty of Humanities and Social Sciences, Prince of Songkla University, Pattani, Thailand, ruthaychonnee.s@psu.ac.th

Sakari Luukkainen, Aalto University, School of Science, Aalto, Finland, sakari.luukkainen@aalto.f

Sandro Battisti, Bruno Kessler Foundation, Center for Information and Communication Technology, Trento, Italy, s.battisti@fbk.eu

Shu-Yu Yeh, Graduate Institute of Management Science, Ming Chuan University Taipei, Taiwan, Republic of China, yehshuu@yahoo.com.tw

Shwu-Ing Wu, Department of Business Administration, National Chin-Yi University of Technology, Taiwan, Republic of China, wusi@ncut.edu.tw

Thakerng Wongsirichot, Information and Communication Technology, Faculty of Science, Prince of Songkla University, Songkhla, Thailand, thakerng.w@gmail.com

Thommie Burström, Hanken School of Economics, Helsinki, Finland, thommie.burstrom@hanken.fi

Yaowalak Trakulmaykee, Human Resources Management, Hatyai Business School, Hatyai University, Songkhla, Thailand, yaowalak@hu.ac.th

Contents

Introduction

Our daily lives are highly influenced by more and more technologies these days. There is no doubt that one of the key technologies is mobile devices. It started with cell phones (or mobile phones, as they are known in some regions of the world), with the basic Global System for Mobile Communications (GSM) in the early 1990s. With this technology, only basic services like telephony were possible. With extensions and further developments, such as the General Packet Radio Service (GPRS) or Enhanced Data Rates for GSM Evolution (EDGE), the first basic internet service could also be provided (or what we know as a "2G" network). The real big step happened through the introduction of the Universal Mobile Telecommunications System (UMTS) (which established "3G" networks as the norm), followed by the more recent introduction of Long Term Evolution (LTE) communication (known as the "4G" network). And we now have the next generation (the "5G") of mobile communication networks just around the corner!

It is a challenge for Information Technology (IT) companies — the producers of smartphones, the network providers — to manage these transitions. They have to invest large amounts of money to setup new or updated infrastructures. In addition, it is also a challenge for customers to adapt and use new technologies. This is especially the case in recent years, where technologies emerge much faster than in earlier days (see Figure 1).

To address these managerial challenges, the book is organized into three parts. First, major technical constructs, which are key in the management of mobile technologies, are introduced in Part 1. Then, an international perspective with a focus on

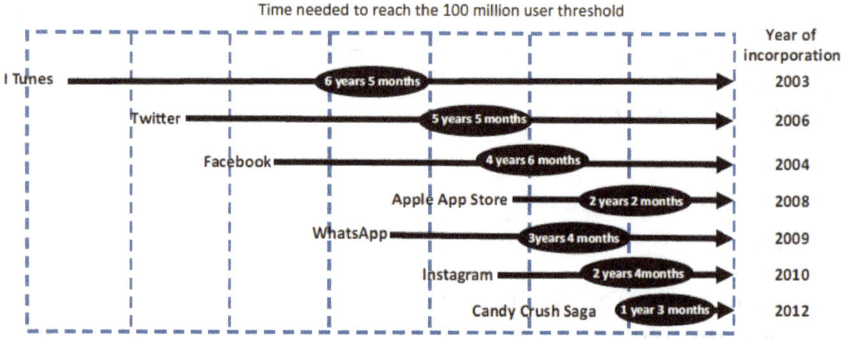

Figure 1. Timeframes to reach 100 million users [Brem and Viardot, 2017, p. 7].

organizational challenges is given in Part 2. And finally, emerging applications for the individual are presented in Part 3.

In Part 1, "Major Technical Constructs", the first chapter introduces an important managerial challenge, which is the ecosystem for mobile technologies. Juha Winter, Sandro Battisti, Thommie Burström and Sakari Luukkainen identify platform constructs and platform leadership as key requisites of business ecosystems; based on that, they discuss success factors of ecosystems and their related platforms in the mobile industry. The following chapter by Chia-Liang Hung and Jerome Chih-Lung Chou focuses on the identification of the cultural influences on the adoption of mobile commerce based on the comparative cases of Taiwan and Malaysia, so as to give insights to mobile operators' global entries. The following chapter presents research on the role of brand awareness. This area is of growing importance, since brand awareness is supposed to have a key influence on the success of new product developments. Hence, brands are treated as another major construct for this book.

Part 2, "International Perspectives and Organizational Challenges" opens with Chapter 4 taking an international perspective with a focus on the Indian market, as India is one of the few markets which still has remarkable growth rates. The chapter sheds light on India's development and success factors. Next, a comparative view on the European mobile market is given in Chapter 5. Here, the success of i-mode is analyzed, and discussed in a European context. Chapter 6 analyzes the impact of mobile phone use on individuals' Quality of Life in Japan. Chapter 7 offers a glimpse into Germany and South-Africa's market developments. The goal of this study is to research the applicability and invariance of the research model in studies in these two countries, whereby the focus is on mobile payment solutions.

In Part 3, "Emerging Personal-level Applications", Chapter 8 discusses the area of emerging applications, the case here being mobile food information. This continues in the succeeding chapter, where the focus shifts to the use of mobile information technology for tourism and leisure. Another important area for emerging technologies is their use in supporting disabled people in their daily lives. Here, the Chapter 9 offers interesting insights into the relationships between the perceived usefulness of mobile phones and user satisfaction for persons with disabilities. Chapter 10 analyzes the role of mobile phones in public places, to study the impact of demographic factors on attitudes that shape the social acceptability of mobile phones, and how this attitude in turn affects the usage of mobile phones. And finally, Chapter 11 deals with the emerging topic of cyberbullying. This chapter analyzes whether a mobile cyberbullying support application can provide useful functions for people affected by cyberbullying.

We hope this book gives a structured overview of mobile technologies to interested readers. As we combine technical constructs with an international perspective, organizational challenges and related emerging technologies, we believe this book also offers a unique insight into the interesting industry of mobile technologies.

Reference

Brem, A. and Viardot, E. (2017). Revolution of innovation management: The digital breakthrough. *Revolution of Innovation Management*, Eds. A. Brem and E. Viardot. Palgrave Macmillan, London.

Chapter 1

Major Constructs: Ecosystems

Juha Winter, Sandro Battisti, Thommie Burström, and Sakari Luukkainen

1. Introduction

The deterioration of Nokia's market share in smartphones, following its renewed strategy in 2011, illustrates a rare event in business history: an industry leader collapses in terms of market share and profitability in just a few years [Canalys, 2014]. Almost simultaneously, however, two new entrants to the smartphone market, Google Android and Apple, have asserted their dominance. Explaining this dramatic change in the market structure has to do with business ecosystems, as explored by Iansiti and Levien [2004]. More specifically, this paper proposes that certain key characteristics or *success factors* (SFs) of the competing business ecosystems, possessed by Google and Apple, led to a chain of events that changed the smartphone industry landscape for good, and that the explicit identification of these SFs deserves special attention.

Nevertheless, these SFs are not easily identified. The literatures discussing potential SFs are scattered and in-depth holistic case descriptions are lacking in this field of research. What is clear is that the success behind mobile ecosystems is related to the *platform* construct. Gawer and Cusumano [2008] identified four *levers of platform leadership* as well as generic strategies for becoming a platform leader. Other platform researchers propose that management of competitive dynamics and pricing are significant SFs [e.g., Rochet and Tirole, 2003]. Indeed, some researchers have also taken a very narrow perspective on the Android ecosystem, focusing on design aspects and user acceptance of technology [e.g., Bouwman *et al.*, 2014], as well as the creation of mobile ecosystems that enable users to control their experiences during the use of technology [e.g., Remneland-Wikhamn *et al.*, 2011; Battisti (2019)]. Still, although both Gawer [2014] and Thomas *et al.* [2014] have tried to bridge various platform concepts, the views on SFs are far from unified.

From this perspective, the main research gap is to explicitly identify and discuss SFs of ecosystems and platforms in the mobile industry, in particular, extending the research of Gawer and Cusumano [2014] and the work of Davis [2016] by analyzing the business dynamics inside the Android ecosystem.

First, management scholars have studied the origins of competitive advantage of companies, in order to explain business performance [Rumelt *et al.*, 1991]. Dyer and Singh [1998] argued that competitive advantage could stem from *idiosyncratic interfirm linkages*, forming a *relational view*. Thus, four sources are defined: (i) *investments in relation-specific assets*, (ii) *substantial knowledge exchange*, (iii) *complementary resources/capabilities enabling the joint creation of new products, services, or technologies*, and (iv) *effective governance mechanisms resulting in lower transaction costs*. Although not explicitly mentioning ecosystems, the above findings are noticeably similar to those of, e.g., Williamson and De Meyer [2012] who identified the "six keys to ecosystem advantage" as follows: *structuring differentiated partner roles*, *enabling flexibility and co-learning*, *stimulating complementary partner investments*, *reducing transaction costs*, *pinpointing the added value*, and *engineering value capture mechanisms*. Based on the above, it has been proposed in this chapter that SFs of ecosystems could well originate from *relationships between firms* and should therefore be further studied.

Thomas and Autio [2012, 2014] define a business ecosystem as a "*network of interconnected organisations, organised around a focal firm or a platform and incorporating both production and use side participants"*, having three common characteristics: a network of participants, a governance system, and a shared logic. The ecological analogy of *business ecosystem* has been analyzed from several different perspectives in innovation and management research, often describing the advances in information and communications technology (ICT). It has been mainly used to explain the competition and collaboration among organizations. From this point of view, the research is focused on exploring the perspective of creating value, in particular, oriented to the creation of mobile business ecosystems. Furthermore, SFs could stem from health measures of business ecosystems, as defined by Iansiti and Levien [2004] and quantitatively by Den Hartigh *et al.* [2006]. Additionally, Bosch [2009] outlines SFs for OS-centric software ecosystems. A key point is to understand that, in innovation ecosystems, innovation is performed through multipartner innovation efforts, but such boundary conditions are understudied [Davis, 2016, p. 33]. This type of literature will therefore be used in a later stage of this paper in order to identify and discuss SFs.

Second, this study uses platform literature in order to identify SFs. Particularly, it draws on Gawer and Cusumano (2008, pp. 2–3), and defines platform as an evolving system made of interdependent pieces that can each be innovated upon. The general literature on platforms can be seen as being divided into four distinct streams of research: (1) *organizational platforms* (with dynamic capabilities as the key construct), (2) *product family platforms*, (3) *market intermediary platforms* (serving two- or multi-sided markets), and (4) *platform ecosystems*, as discussed by Thomas *et al.* [2014]. This is compatible with and builds upon the typology of platforms by Gawer [2009], comprising *internal platforms*, *supply chain platforms*, *two-sided (or multi-sided) markets*, and *industry platforms*. In our study of ecosystem SFs, particularly, the last two streams of research are of relevance and will be further discussed below.

Multiple researchers emphasize the importance of understanding the logic behind *two-sided markets* [Parker and Van

Alstyne, 2005; Rochet and Tirole, 2003; Eisenmann *et al*., 2006]. Within this line of research, it is further proposed that the issue of pricing [e.g., Hagiu, 2006, 2009b] is an important topic, and so is the subsidization of the two-sided markets whose transactions are facilitated by a platform [Armstrong, 2006; Armstrong and Wright, 2007; Hagiu and Spulber, 2013]. Moreover, Hagiu [2009a] and Rysman [2009] discuss that *multi-sided platforms* could be considered as a generalization of two-sided markets. Although case studies are common within this stream of literature, often using credit cards, computer operating systems (OS), or business directories as examples [e.g., Rysman, 2004], relevant studies of mobile ecosystems are lacking.

Although being presented as two different concepts, there is actually a strong resemblance between platform ecosystem studies and industry platform studies. *Platform ecosystems*, as presented by Thomas *et al*. [2014], roughly equate to industry platforms as discussed by Gawer and Cusumano [2008, 2014]. Both streams of research view the focal firm as a hub or a central point of control in technology-based business ecosystems. Contextually, the literature has focused on the information technology and internet sectors, with extensive case studies on, e.g., Intel [Gawer and Henderson, 2007]. Still, these studies can work as inspiration for defining SFs in mobile ecosystems.

A special issue to consider for ecosystem leaders in mobile ecosystems is the question of openness [cf. Eisenmann, 2008]. Thomas *et al*. [2014] elaborate on this issue and present a model of *architectural leverage*, combining three distinct logics of leverage, namely, *production*, *innovation*, and *transaction logic*, with *architectural openness* based on the degree of disaggregation of the industry value chain, also identifying situated types of platform variants within the model. The authors further argued that market intermediary platforms and platform ecosystems typically exhibit so-called *many-to-many architectures* where both supply and demand sides are open to participants. This seems compatible with Eisenmann's [2008] dimensions of platform openness, although he defines openness also in terms of two additional roles, the platform provider and sponsor.

Furthermore, Muegge [2013] made an attempt to integrate the theories of industrial platforms and business ecosystems, arguing that, in a complex hierarchical system, they represent different but complementary layers of analysis. In a similar line of argumentation, Gawer [2014] observes the resemblances between roles such as being a *platform leader* and/or a *keystone firm.* Thus, both Muegge and Gawer communicate the importance of the possibility to use both the firm level and the ecosystem level as distinct levels of analysis.

Thomas *et al.* [2014] also identified different levels of analysis but did not explore the theory of platform concepts beyond the level of the firm. They did allude to, e.g., industry- and ecosystem-level analysis and advocated further research from different analytical perspectives. Thus, it would appear that, in management research, there is interest in multilevel analytical frameworks and more holistic exploration of the theory behind platforms and ecosystems. Furthermore, although Gawer's [2014] proposal may well be the most advanced integrative framework of platform concepts to date, she does not discuss the SFs of platforms or ecosystems, nor the interaction of decisions related to different parts of the framework such as platform scope (including possible envelopment) and openness. Also, as she acknowledges herself, the framework requires systematic empirical validation.

Finally, indeed some scholars have explored the level of analysis of software ecosystems and developer marketplaces with particular focus on those of Apple, Google, BlackBerry, and Microsoft [see, e.g., Campbell and Ahmed, 2011; Idu *et al.*, 2011; Tuunainen *et al.*, 2011; Hyrynsalmi *et al.*, 2012]. In particular, the main contributions of these scholars have been in the area of application marketplaces and developer programs. However, these researchers have lacked a holistic understanding of the platforms and methods for capturing value inside the respective ecosystems of organizations. Therefore, in Sections 1.1

and 1.2, the particularities of the firm-level and ecosystem-level analysis dimensions are discussed.

1.1. *Firm-level dimension*

Firm level forms the first and most granular part of our framework, since we look at the value creation and extraction of the platform leader or ecosystem orchestrator, being the central firm or *keystone* of its ecosystem, followed by a discussion of the firm's platform approach and governance. It is related to the breadth of business activities of companies such as the product offering, development, and commercialization, as well as the services delivered to specific customers and the place where the business activity is carried out.

Various monetization models may be mixed and utilized by firms sometimes simultaneously. In our analysis framework, the method of value configuration analysis as defined by Stabell and Fjeldstad [1998] is applied for the purpose of identifying alternative means of value creation, serving as a generalization of traditional value chain analysis. In spite of this, our study considers the question of monetization reflecting on existing capabilities and assets, also considering historical developments. The theory of path dependence [David, 1985; Arthur, 1989] supports our idea that the historical legacy of a firm mainly in terms of its previous business and related activities, capabilities, and assets has had an impact on more recent strategic decisions and choices the firm has made with regard to ecosystems and platforms.

The *four levers of platform leadership* [Gawer and Cusumano, 2008] and the descriptions of the roles of platform leaders and complementors form the basis for the study of the case company's platforms. We separate the analysis of how much the firm innovates and produces complements in-house from how the firm manages and incentives its external complementors. The latter point is also relevant in the analysis of ecosystems, where software ecosystems and particularly application marketplaces are examined. We also consider the strategic approaches taken by firms to become platform leaders such as "coring", "tipping" [Gawer and Cusumano, 2008], expanding the scope of one's platform by incorporating some of the functionality of the platforms being used in an adjacent market.

The control paradigm of a platform may be either *proprietary* or *shared*, and this has profound implications especially on value capture and the management of complementary innovation. Eisenmann [2008] argued that both paradigms can be successful and examines factors that are favorable to each when designing new platforms. Shared platforms are prone to free rider problems that make it challenging to protect returns from infrastructure investments or to offer user subsidies. In contrast, proprietary platforms do not have such problems.

Regarding the effects of the number of complementors on the intensity of competition, Boudreau [2008] argued that having too many complementors can reduce a platform's ability to generate new innovation and profits for firms making complements on it, mainly due to crowding-out effects and substitution instead of market expansion. These effects would both affect the attractiveness of the platform negatively from end-user and complement the producer perspective.

Differentiation through either vertical integration or exclusivity is often sought by platform providers to make themselves stand out from their rivals. Lee [2013], who has studied this phenomenon and its implications, contended that exclusive software is a key leverage for platform entrants, allowing them to differentiate and gain market traction more effectively. In terms of the overall market, however, exclusivity actually decreases the total market revenue as well as consumer welfare.

This research focuses mostly on technical and organizational concepts, comprising mainly analysis on firm level. While network effects and multi-sided markets do play a fundamental role also in platform theory, we discuss the related analysis dimensions under *ecosystem-level analysis*.

1.2. *Ecosystem-level dimension*

Iansiti and Levien [2004] have studied the characteristics of business ecosystems such as their structure and health. The three critical measures defined by them, *productivity*, *robustness*, and *niche creation*, are also applied in our analysis framework. Alternative health metrics have been suggested by Den Hartigh *et al.* [2006], divided into partner health and network health metrics. These metrics, as they argue, are better suited for measuring the health of business ecosystems on the company level and of more practical value to managers.

In our framework, the *software ecosystems* considered are OS-centric in the sense that the OS and its interfaces play a key role in defining what really constitutes the software ecosystem. They also set the technical boundaries for what kinds of complements can be produced on a particular platform. We use the following SFs identified by Bosch [2009] specifically for OS-centric ecosystems for the evaluation of the case company's software ecosystems: (i) *minimal effort required by developers to build applications on top of the OS, thereby enabling both breadth and quality of the application offering*, (ii) *generic, evolving functionality and set of features provided by the OS that maintains attractiveness for developers*, and (iii) *the number of customers that use the OS and that are accessible to developers for monetization*.

Being a key component of a mobile software ecosystem, a mobile *application marketplace* is also a well-known example of a two-sided market, and thus relevant characteristics of two-sided markets are used in the case study to analyze them. Joining multiple platforms to make one's products available on more than one market, i.e., multihoming, is a key phenomenon in mobile application stores as discussed by Hyrynsalmi *et al.* [2012].

The *openness* of a platform ecosystem is defined by Eisenmann *et al.* [2008] along four distinct dimensions, each of which corresponds to a certain role in a platform-mediated network: demand-side platform users, supply-side platform users, platform providers who operate the platform and interface directly with the customer, and platform sponsors who own the platform and decide who gets to participate. Many kinds of platforms can be analyzed using this set of dimensions of openness, not tied to a particular industry. However, specific to mobile application stores, Müller *et al.* [2011] identified a total of 12 distinct value network roles, namely, *end user, network operator, payment broker, advertisement broker, marketplace, OS developer, testing and verification party, signing partner, software developer, content provider, software distributor*, and *device manufacturer*. Our framework applies this extended list to evaluate the openness of the platforms and ecosystems.

2. Research Design

The research methodology is a case study, following an in-depth empirical inquiry of a contemporary phenomenon as proposed by Yin [2003]. We focused on analyzing a specific case in depth, aiming to identify relevant SFs of the case and not to generalize the findings of the case. Our study is primarily based on a literature review of both academic (i.e., using the Scopus database and Google Scholar) and non-academic literature, following the secondary data analysis method proposed by Heaton [2008]. Furthermore, while success could be defined in a number of ways, we considered SFs that enable innovations to be diffused throughout the markets (e.g., Rogers [2003]).

The main motivation for the case selection is that the Android ecosystem is a very particular success case in the literature and in managerial practices. In particular, the justification for selecting this mobile business ecosystem for our case study is that in just six years between 2008 and 2014, it has been able to capture an unparalleled 80% share of the global smartphone market [Canalys, 2014].

Because the subject matter of the study is very topical and developing rapidly, receiving up-to-date information required the extensive utilization of non-academic literature such as magazine articles and web pages related to the Android ecosystem, the collection of which was based on online sources published between June 2005 and December 2015. These literary sources were carefully selected based on their accurate descriptive nature, providing detailed evidence and insights about the subject matter so that we could apply the analysis framework created for this study. Moreover, we carried out the analysis without using any special software tools, because we considered it unnecessary for the proper understanding of SFs in business ecosystems.

We have created an analysis framework for the empirical part of this study. As recommended by Davis [2016], this paper moves beyond the dyad and instead studies complex boundary conditions. The main goal of our framework is to explore both *firm*-level (including internal and supply-chain platforms) and *ecosystem*-level (including industry platforms) SFs of mobile business ecosystems as discussed in the previous section. When discussing platforms, we do not make a clear division between supply-chain platforms and industry platforms, as a certain degree of architectural openness exist in both cases.

Following the criteria presented above and in the theoretical background, we have synthesized a set of metrics representing the various dimensions of analysis that we utilize in this research, as presented in Table 1. For each metric, a unique identifier (e.g., **F2.1**) is assigned, that is used later in this paper whenever referring to that particular metric. Furthermore, the key references pertaining to each metric are listed after the description.

Table 1. Analysis framework dimensions.

Levels	Dimensions	Metrics
Firm	**F1**. Firm scope and value creation	**F1.1**. Understanding the sources of revenue, also from a historical and path-dependent perspective [David, 1985; Arthur, 1989]
		F1.2. Design of the monetization models and the value configuration of the firm [Stabell and Fjeldstad, 1998]
	F2. Platform approach and governance	**F2.1**. Extensibility and openness of platform architecture and technology [Gawer and Cusumano, 2008; Gawer, 2014]
		F2.2. In-house versus external focus in complements [Gawer and Cusumano, 2008]
		F2.3. Managing and incentivizing complementors [Gawer and Cusumano, 2008]
		F2.4. Internal organization and propensity to advance the overall good of the ecosystem [Gawer and Cusumano, 2008]
		F2.5. Control paradigm/openness [Eisenmann, 2008; Eisenmann *et al.*, 2008]
		F2.6. Vertical integration and exclusivity [Lee, 2013]
Ecosystem	**E1**. Ecosystem approach and governance	**E1.1**. Role of the firm in the ecosystem [Iansiti and Levien, 2004]
		E1.2. Ecosystem health metrics (e.g., productivity, robustness, niche creation) [Iansiti and Levien, 2004; Den Hartigh *et al.*, 2006]
	E2. Software ecosystem and application marketplace	**E2.1**. Occurrence of SFs for OS-centric software ecosystems [Bosch, 2009]
		E2.2. Openness along various roles in the value network: end users, developers, platform providers, platform sponsors [Müller *et al.*, 2011]
		E2.3. Occurrence of multihoming [Hyrynsalmi *et al.*, 2012]

3. Data Analysis and Propositions

This section serves the purpose of analyzing the SFs of the Google Android ecosystem. We use a step-by-step approach where we first make an analysis on the firm level, followed by an analysis on the ecosystem level, according to the framework and the metrics presented in Table 1. Whenever appropriate, each paragraph is linked to a specific metric or metrics as per Table 1, that the evidence presented is linked to. This is denoted after each paragraph in parentheses.

3.1. *Firm-level analysis*

Google's advertising business model is by and large built on two programs: AdWords and AdSense. However, Google also provides services such as display advertising, Google Apps, and Google Drive. Some of these services are subscription-based and some are built on a freemium model. That is, customers can use, for example, search services free of charge as long as they can be targeted for ads and promotion. For sure, Google also sells some hardware such as Chromebooks and Nexus/Pixel products. Thus, Google's primary revenue model can be described as sales of search and display advertisements on the web and on mobile devices, which is complemented by sales of subscriptions for cloud-based apps and services. Nevertheless, the main source of income is still based on sales of advertisement, apps, and other services (*metrics **F1.1** and **F1.2***).

More importantly, Google is also renewing its business model. Google is using its close relationship with the Open Automotive Alliance (OAA) in order to establish itself in the automotive industry. This strategic move can be realized through close collaboration with the already established manufacturers in the automotive industry. This type of strategic work shows some similarities with Google's previous strategy of connecting to the smart home and home entertainment ecosystem, the automotive ecosystem being yet another area of life where

Google desires to be present. Google has also been testing a "self-driving car". The idea is to create a fully autonomous car navigating using sensors and highly accurate data, which can be also used to improve pedestrial traffic safety (e.g. Luukkainen *et al.* 2016). Just as in an airplane, the driver would have more of a supervising role (*metric* ***F1.2***).

Cloud-based services have become a central part of the Google business model. The cloud-based marketplace labeled Google Play has well over 1 million applications and games in its catalogue in addition to other media content. The company has also introduced a virtual wallet service (Google Wallet). This service offers secure management of debit, credit, and reward cards. Google also offers a cloud-based storage service directed toward consumers and small businesses (*metric* ***F1.2***).

Google has reduced the amount of complexity by focusing on software and cloud services while supporting largely generic hardware. Through this strategy, people who access the Internet are likely to use Google services, or Google Network Members' websites, and consequently such use generates income for Google (*metrics* ***F1.1*** *and* ***F1.2***).

Google has, through its strategy, created a high brand value; the brand is obviously very appreciated by its customers. Google was ranked as the third most loved firm according to Sustainable Brands [2013] and reached a fifth place in relation to brand value [Forbes, 2013]. A high brand value, in turn, contributes to demand for the company's services, strengthening its ecosystem. Thus, we propose the following SF:

- **SF1**. If a platform-agnostic business model is used by a company, with independence regarding types of devices and hardware supported, then the likelihood of success of the company's business ecosystem is increased (*metrics* ***F1.1*** *and* ***F1.2***).
- **SF2**. If no burden of legacy in the form of a hardware business exists at a company, which enables a clear focus on cloud services, software, and content, then it increases the

likelihood of success of the company's business ecosystem (*metric **F1.1***).

- **SF3**. If a company and its services are loved by consumers, combined with a high level of brand equity, then it increases the likelihood of success of the company's business ecosystem (*metric **F1.1***).

With more than a billion Android devices in use globally as of 2014 [Gartner, 2014], Google has reached a very competitive position such that the firm can make use of the swift growth in the smartphone industry and the huge number of connected devices running on Android. Although Google is not monetizing Android directly, it can still benefit from knowledge about customer preferences, their online activities, and patterns of usage. Through this knowledge, Google can build user profiles and have a high precision in targeted advertising (*metric **F1.1***).

In terms of strategy, Google avoids vertical integration and physical domination in the smartphone industry. The Google strategy is to invite complementors and device manufacturers to innovate on top of the Google platform [Google, 2014]. The logic behind this strategy is simple. Google strives to promote the proliferation of Android and therefore encourages the creation of external devices and apps that are Android compatible (*metrics **F2.2**, **F2.3**, and **F2.6***).

Android was acquired by Google in 2005, and gained customer acknowledgement after Google announcing the creation of the Open Handset Alliance (OHA), using this alliance as a way to promote Android. Through this action, Android was positioned as an open and comprehensive mobile software platform. The openness of the platform is further strengthened through the Android Open Source Platform project (AOSP), as argued by Pon *et al.* [2014], thus (*metric **F2.5***).

The Google strategy has been to develop future versions of Android, and its core platform application programming interfaces (APIs), through a more closed development branch. Such development has usually also been done in collaboration with an original equipment manufacturer (OEM). This strategy can be seen as a deviation from the open source strategy. However, Google claims that it is in the best interest of all stakeholders, since it allows for developers and OEMs to focus their development on readymade versions of the platform. This is not in line with the wishes of most OEMs, as they typically want to access the latest version of the platform. For the developers, however, this strategy fits well since they are interested in applying their solutions to a stable platform. Thus, we propose the following SFs (*metrics **F2.1**, **F2.2**, **F2.3**, **F2.4**, and **F2.5***):

- **SF4**. If a company gives more freedom to complementors with regard to developing service innovations on top of the company's technologies, then it increases the likelihood of success of the company's business ecosystem (*metrics **F2.2**, **F2.3**, and **F2.5***).
- **SF5**. If a platform exposes enough APIs so that OEMs, accessory makers, and developers are able to create complementary products and apps with meaningful differentiation, then it increases the likelihood of success of the platform's business ecosystem (*metrics **F2.1** and **F2.2***).

Although Google is taking some steps to open up its core technologies also with Android, the fact remains that Google mobile services (GMS), Google Play services, and a multitude of other Google's applications and services suite remain proprietary and protected by the company's intellectual property rights. These services and products will not be part of AOSP, i.e., not part of open source [InfoWorld, 2014]. OEMs have also been forced to follow procedures such as the "GMS approval window". This measure has pushed OEMs to deliver products not older than nine months if they want to keep using the Google apps and services [DailyTech, 2014], thus (*metric **F2.5***).

Taking this approach, Google performs a strategy of proprietary control in relation to Android platform key elements. These elements should be seen as critical for the platform.

Furthermore, most device manufacturers do not create and ship products using only the AOSP code. Instead, they use the proprietary GMS suite licensed from Google. The behavior of relying on GMS can be explained by the fact that for example Google Play, Drive, Maps, and Gmail-related APIs, central for application development, are also part of GMS. Consequently, application developers too have to decide on targeting the lean AOSP platform (with fewer APIs) or the more extensive GMS-equipped Android devices [InfoWorld, 2014]. The flipside for developers using the GMS solution is that it comes with multiple requirements giving Google increased control of Android-based devices [InfoWorld, 2014]. Thus, we suggest the following SF (*metrics* ***F2.3*** *and* ***F2.5***):

- **SF6**. If the software components of a platform that meaningfully differentiate it from the competition or otherwise create significant added value that are kept closed source, then it increases the likelihood of success of the platform's business ecosystem (*metric* ***F2.5***).

External software development has been a possibility to the general community since the Android platform was announced in November 2007. Software developers can access the Android platform's developer resources through a unique developer site. Developers can, without having to join a developer program, test their software and use the Android development tools free of charge. This is not the case with competing platforms such as Apple's iOS platform (*metrics* ***F2.1*** *and* ***F2.3***).

Google has nurtured Android applications development by announcing two Android Developer Challenges. The first challenge was announced in January 2008 and the second in May 2009. An impressive total of $10 million was awarded to developers. The first challenge was of significant importance in order to jump-start the creation of applications, and there were almost 2,000 contributions from over 70 countries. The second challenge helped in creating a critical mass of applications so that the platform became big enough to attract developers without any further nurturing. Moreover, it is supported by the research of Battisti [2012, 2014] and Magnusson *et al.* [2016] on the social aspects of user involvement (*metric **F2.3***).

Google sensibly implemented a similar revenue sharing model as Apple on the application marketplace. However, the publishing process was less rigid. Applying a method with a lack of rigorous pre-inspection before publication has a downside. Therefore, Android has had to deal with inappropriate software in a much bigger scale compared to competitors as iOS and Windows. However, in 2012, Google took measures and installed an in-house automated antivirus program labeled "Bouncer". Installing that program resulted in a reduction of malicious apps with 40% [SC Magazine, 2012]. Furthermore, Google has also a reputation of allowing third-party application even if they should overlap with Google offerings, something that contrasts the policies of, e.g., Apple. Thus, we suggest the following SFs (*metrics **F2.3**, **F2.5**, and **F2.6***):

- **SF7**. If a platform is open with regard to end users, developers, platform providers, and platform sponsors alike, it enables a high degree of open innovation but possibly also major fragmentation, nevertheless increasing the likelihood of success of the platform's business ecosystem (*metric **F2.5***).
- **SF8**. If a non-exclusive process of application development, verification, and publishing is applied on a platform, then it increases the likelihood of success of the platform's business ecosystem (*metrics **F2.3** and **F2.6***).
- **SF9**. If a platform has the most non-restrictive policy for complementors, not limited by protective clauses against competition or substitution, then it increases the likelihood of success of the platform's business ecosystem (*metrics **F2.5** and **F2.6***).

The analysis above gives light to some critical firm-level SFs. Still, this picture of SFs is incomplete. An ecosystem-level analysis is therefore performed in the following section.

3.2. *Ecosystem-level analysis*

In this section, we cover the ecosystem approach and governance of Android, Google's role in it, and the overall health of the ecosystem as well as the software ecosystem and application marketplace.

Google has built an ecosystem around Android where participation should be built on voluntary action. The idea of voluntarism is closely related to the idea of using open source code, as in AOSP, where anyone can change, use, and distribute the software to meet unique needs. Android product development teams typically focus on a limited number of devices and then integrate their hardware with the latest development version of the Android software so that new product launches can be secured, supporting the roll-out of the latest Android version [9to5Google, 2011]. This strategy using so-called flagship devices has the benefit of also carrying the risk for the wider OEM community. This community can make follow-up device development that takes advantage of new features. Google managers claim that this strategy ensures coevolution between the platform and the need for contemporary functional devices. Nevertheless, Google is not showing consistency in its Android strategy. They have been rotating partners for different releases, with varying terms of exclusivity. Thus, we suggest the following SFs (*metrics **E1.1** and **E2.1***):

- **SF10**. If the OEMs benefit from having a state-of-the-art mobile OS platform, essentially royalty-free and enjoying higher adoption rates than any competing platform, then it increases the likelihood of success of the platform's business ecosystem (*metric **E1.1***).
- **SF11**. If generic, evolving functionality and a set of certain key features provided by the OS are available on the platform, then it increases the likelihood of success of the platform's business ecosystem (*metric **E2.1***).

In June 2014, Google made an announcement; the program called "Android One" was presented. The program was targeted at the lower end of the smartphone market. In collaboration with suppliers and OEMs, Google had defined a reference platform for devices that would meet a certain minimum standard of technical specifications. These products would be able to support Google's mobile services in their entirety. Yet, the price would be kept at around US$100, then a modest price for such devices [BBC News, 2014]. The program would ultimately lead to improved performance of low-priced products and also improve the user experience of these products. Previously, the user experience of such products had been somewhat disappointing. In 2016, the program was extended to mid-range smartphones and new, developed markets such as Japan, totaling 21 countries [Google, 2016].

The Google management team is obviously aiming to secure a bigger chunk of the entry-level smartphone market, which is growing faster than any other handset segment. They are at the same time trying to unify the Android experience with standardized hardware and limit fragmentation by barring OEM skinning of the Android One product. Clearly, this initiative played a key role in reaching the large number of prospective smartphone users in emerging markets. The program allowed for reaching these users with Android and Google services in a controlled fashion. As a consequence, Google has also secured a fast-growing segment for its mobile advertising business. This step is vital in order to ensure future growth. With other smartphone OS providers such as Microsoft and other initiatives such as Firefox OS aiming to utilize the same opportunities in emerging markets, Google is seeking to secure its leading position by taking proactive steps. Thus, we propose the following SFs (*metrics **E1.2** and **E2.1***):

- **SF12**. If mobile operators benefit from the transition to smartphones, driving up demand for mobile data plans, then it increases the likelihood of success of mobile business ecosystems enabling affordable smartphones (*metric **E1.2***).

- **SF13**. If the number of customers that use the OS platform and that are accessible to developers is high compared to other platforms, then it increases the likelihood of success of the platform's business ecosystem (*metric **E2.1***).

Using the definition of Iansiti and Levien [2004], Google can be defined as a *keystone* — a platform provider who enables and facilitates value creation for the whole of its ecosystem. In the Google ecosystem, website owners generate income by hosting Google's advertisements, and advertisers make use of Google's broad customer reach and effective high-converting, targeted ads. Some other stakeholders such as handset accessory makers appreciate the possibility to work with standard, well-defined interfaces without being dependent on any single OEM. The community of developers benefit from accessing the broad installed base of Android device owners, giving them the possibility to develop commercial or free apps. Developers additionally appreciate the open and non-restrictive approach of Android in terms of app distribution (*metrics **E1.1** and **E2.2***).

Google has organized its business somewhat differently from rivals like Apple. As a matter of fact, Google earns only very limited revenue from selling small volumes of devices and it does not charge for Android licenses. Unlike Apple, Google has refrained from utilizing proprietary interfaces on its devices and has not introduced such to Android either. Additionally, although multihoming is a relatively rare phenomenon, the Android ecosystem appears to benefit from it more than, e.g., the Windows Phone ecosystem [Hyrynsalmi *et al.*, 2012]. Consequently, Google can be seen as benevolent, willing to provide possibilities for its ecosystem members to earn revenue and grow their business as they see fit. Thus, we propose the following SFs (*metrics **E2.2** and **E2.3***):

- **SF14**. If accessory makers appreciate that they can work with standard, well-defined interfaces, then it increases the likelihood of success of mobile business ecosystems and platforms that support such standard interfaces (*metric **E2.2***).

- **SF15**. If developers value the openness and non-restrictive philosophy in application distribution, then it increases the likelihood of success of mobile business ecosystems that exhibit such policies (*metrics **E2.2** and **E2.3***).

There are also some health issues in the Android ecosystem that are of interest to discuss. For example, the Google Play marketplace offered just about 1.5 million Android applications in its catalogue as of January 2015 [AppBrain, 2015]. Such a vast offering of apps is clearly a sign of healthy diversity that we consider as an adequate proxy for niche creation, and also a sign of productivity. Actually, already in late 2012, Android managed to effectively close the gap in terms of available applications compared to iOS (see, e.g., Hyrynsalmi *et al.* [2012]). This, in addition to additional indicators such as where applications are first published and how soon they migrate to other ecosystems, could provide additional insights on ecosystem's health and success.

It is unfortunately very difficult to evaluate ecosystem robustness since there are no statistics illustrating the survival rates of developers. However, robustness might be questioned since a high proportion of developers only publish one application (58%). The situation also seems to be the same in the iOS ecosystem. This translates to higher churn than with large, professional developers. It should also be taken into account that from Android's fragmentation, it follows that the platform scores less satisfactorily in terms of *limited obsolescence* and *continuity of user experience and use cases*. These are metrics associated with the ecosystem health measure of robustness according to Iansiti and Levien [2004]. Thus, we suggest the following SF:

- **SF16**. If an ecosystem is healthy, based on metrics of productivity, robustness, and niche creation, then it increases the likelihood of success of that ecosystem also in a mobile business context (*metric **E1.2***).

4. Discussions and Findings

In order to present the main contribution of this research, a model that helps to analyze the SFs of mobile ecosystems, we grouped the 16 factors into four findings, presented as follows:

- **Finding 1**. Google's platform-agnostic business model, independent of devices and hardware and free of any burden of legacy, enables a clear focus on cloud services, software, and content, which in turn has enabled Google as well as its Android ecosystem partners to thrive (SF1 and SF2).

The first finding elaborates the notion that Google's firm scope, its historical legacy, and current business and monetization model are beneficial to the Android ecosystem in the sense that they leave plenty of room and ways for ecosystem members to create value. Search advertising remains Google's key revenue source and, ultimately, any devices or platforms can serve to drive this revenue, not just Android. Unlike the orchestrators of some other mobile business ecosystems, Google does not attempt to capture the majority of the value created in the ecosystem, a behavior favored by physical and particularly value dominators, as described by Iansiti and Levien [2004]. Moreover, as the company lacks a burden of legacy and path dependency in its technology choices, it has been able to avoid being locked in to unfavorable technologies, also allowing its ecosystem partners more freedom to innovate and produce complements. This finding also appears to support the third hypothesis of Gawer [2014] stating that "*collaborative governance will increase complementors' incentives to innovate in platform-enhancing ways*", although whether all innovation in the Android ecosystem is platform-enhancing is subject to debate.

- **Finding 2**. While the Android platform has a generic, evolving set of functionalities and exposes enough APIs so that various complementors are able to create products and apps with meaningful differentiation, the software components of the platform that meaningfully differentiate it from the competition or otherwise create significant added value that are kept closed source (SF5, SF6, and SF11).

With the second finding, we emphasize the importance of having a sufficiently broad and evolving set of APIs that enable complementary innovation with potential for meaningful differentiation, a key prerequisite for a viable third-party application ecosystem. This is also in line with the observations of Bosch [2009] for OS-centric software ecosystems. At the same time, however, Android also has certain key components developed by Google that are kept closed source, and it would appear that the number of such components has increased rather than decreased as of late. This is to protect the substantial investments made by Google against free riders as described by, e.g., Eisenmann [2008], and to ensure that certain key features of the platform remain positive differentiators and are not simply copied by competing platform providers. Furthermore, it would appear that Google is also taking action in response to certain complementors-turned-competitors like Apple, who decisively replaced Google Maps with its own mapping software, or Facebook, who exploited Android APIs to create its own UI layer replacing the standard Android home screen, thus directly competing with Google for end-users' attention and advertising revenue. These two examples were also used by Gawer [2014] to justify her second hypothesis (*a number of platform complementors will turn into competitors over time even in platform ecosystems whose participants are largely complementary*) and fourth hypothesis (*emergence of competition from former complementors is likely to elicit a reaction from the platform leader, either closing its technical interface or enveloping the complementors-turned-competitors*).

- **Finding 3**. Android's appropriate balance of architectural openness, evident in mostly open and non-exclusive but also some closed policies, processes, and interfaces with regard to complementors such as device manufacturers, accessory makers, and application developers, content providers and publishers, platform providers such as alternative application store operators and mobile operators, has enabled a high degree of innovation and added value in complements but also potentially severe fragmentation that is already evident in the various forked versions of Android available (SF3, SF4, SF7, SF8, SF9, SF14, and SF15).

The third finding emphasizes the partially shared, partially proprietary control paradigm and considerable openness of the Android platform and ecosystem with regard to the dimensions defined by Eisenmann *et al.* [2008] and Müller *et al.* [2011]. It would not be very meaningful to simply say to that Android is either "open" or "closed" as the matter is more complicated than that, depending on multiple dimensions of openness relating to the value network and ecosystem around the platform. Moreover, as argued by Müller *et al.* [2011], both relatively open and closed models have advantages and disadvantages.

Also, what works for the Android ecosystem may not be ideal for just any ecosystem, as is the case with, e.g., the considerably more closed iOS ecosystem, where Apple as the sole device manufacturer and the platform provider takes as much as 65% of the whole smartphone industry's profits, according to Canaccord Genuity analysts' statistics for the first quarter of 2014, cited by Fortune [2014]. The iOS ecosystem has low fragmentation, whereas with Android, the breadth of releases and forked variants of the platform in use is remarkable. The stricter testing and verification approach of iOS combined with a limited number of different hardware devices supported has contributed to an application catalogue where the number of low-quality applications is generally lower than with Android, the latter having 15% of its applications ranked as low quality according to AppBrain [2015].

Müller *et al.* [2011] argued that the iOS approach might lead to a better user experience and help prevent diseconomies of scale. On the other hand, they argued that the more open approach of Android would support a strategy aimed at becoming the largest application store provider, and it may lead to increased competition in both software and hardware, resulting in lower prices. Regardless of this, both platforms have, broadly speaking, an equal number of applications available, although Android enjoys a much larger share of the smartphone market than iOS, 80% versus 15% for 2013 [Canalys, 2014]. It would seem that Android's model of openness and governance works for a more distributed ecosystem with less vertical integration, whereas the iOS model works for a highly vertically integrated, centralized ecosystem where most roles in the value network are under proprietary control.

Again, this finding could be seen to support Gawer's [2014] third hypothesis but also her first hypothesis, "*as the platform interfaces become more open, more agents will be attracted into the platform ecosystem, and the platform leader will be able to access a larger set of potentially complementary innovative capabilities*". As evidenced by the fragmentation of the Android platform, however, extensive openness can also work against the platform despite spurring a higher degree of complementary innovation.

- **Finding 4**. The strong market position and healthy, established ecosystem of Android appeal to OEMs building devices on the platform, operators selling mobile broadband subscriptions and devices, developers and other complementors making complements for the devices, and ultimately consumers who enjoy an unrivaled variety in Android devices, accessories, and applications (SF10, SF12, SF13, and SF16).

The fourth finding highlights the importance of the Android ecosystem's health to its various members, particularly, in terms of metrics such as productivity, robustness, and niche creation as defined by Iansiti and Levien [2004]. Furthermore, the large established user base of Android device owners, contributing to strong network effects, makes the ecosystem attractive to device OEMs and complementors such as application developers. Accounting for four-fifths of the global smartphone market, Android is simply too large to be ignored by any party willing to address a major demographic of smartphone users. Furthermore, operators in many countries see affordable smartphones based on Android driving the transition from voice-centric feature phones to mobile broadband-enabled smartphones, helping them sell more mobile data plans and at least partially offset the decline in traditional voice and text message revenue that, not coincidentally, is also related to the proliferation of smartphones and certain popular applications and services such as Skype and WhatsApp.

The above four findings together constitute the answer to the research question. To illustrate the relationship between these findings in the context of our multilevel analysis framework, we have crafted a model, as presented in Figure 1.

Although the findings are derived from the case study, the model of SFs itself is not specific to just Android. The basic notion of the model is that as the level of analysis becomes less granular, moving from firm-level analysis through platform analysis to ecosystem-level analysis not discretely but in a continuum as illustrated by the arched arrow, we eventually cover the whole breadth of SFs. Some of these may be attributed just as well to platform characteristics as to ecosystem characteristics, as is the case with Finding 3 that is enclosed both in the horizontal and the vertical dotted-line rectangles that group together ecosystem-related and platform-related SFs, respectively. The lower half of the diagram represents firm- (and platform-) level analysis (dimensions F1 and F2 in Table 1), whereas the upper half represents ecosystem-level analysis (dimensions E1 and E2) with some aspects shared with platform-related analysis (dimension F2). Both points appear compatible with the thinking of Gawer [2014] whose classification

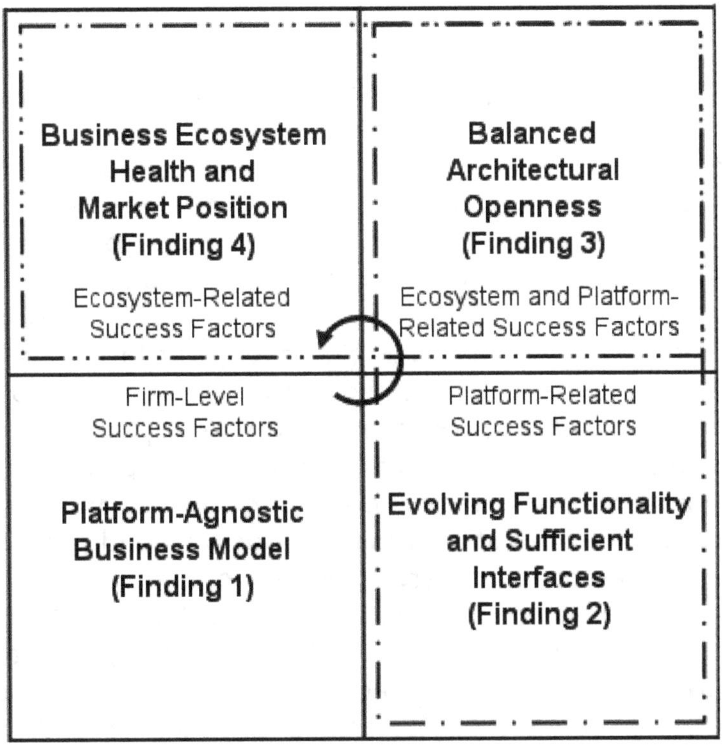

Figure 1. Ecosystem SF model.

of technological platforms is also a continuum where the last category, industry platform, takes the organizational form of an ecosystem. Similarly, Thomas *et al*. [2014] see platform ecosystems as the ultimate platform type that combines many-to-many architectural openness (as in Finding 3) with regard to three distinct logics of leverage: production, innovation, and transaction logic.

5. Conclusions and Implications

Although business ecosystems and platforms have been studied also in the context of the smartphone industry, few existing studies have considered firm-level as well as ecosystem-level aspects and dimensions of analysis jointly, integrating the different but related concepts into a single framework. Thus, this paper draws on Davis [2016] who recommends studies of group dynamics that go beyond the traditional dyad studies traditionally performed in this field of research. The proposed model contributes to the literature of innovation platforms via clustering the SFs leveraged both on firm-level and ecosystem-level analysis, which is the key contribution of this study.

We identified 16 SFs for the Google Android ecosystem, clustered into four findings pertaining to: (i) the business model and focus of the ecosystem orchestrator, (ii) a generic, evolving set of functionality and APIs enabling the creation of differentiating complements, (iii) the balance of mostly open and some closed policies with regard to ecosystem partners and complementors, and (iv) a strong market position and a healthy, established ecosystem. These findings should help scholars of innovation management research as well as managers (particularly those that operate in the smartphone industry or adjacent industries such as consumer electronics or automotive infotainment) understand the key factors that have contributed to the rise and success of mobile business ecosystems. Especially, the key implication for managers could be summarized as: whether companies are seeking to join the Android ecosystem or perhaps replicate its success in a different industry context, they would do well to implement or enable the key SFs in their companies.

Policy makers could enable the development of innovation platforms that create economic and social values. From this perspective, the contribution of our research for policy makers is a conceptual model with critical SFs of mobile business ecosystems, which can be taken into consideration during the definition of initiatives for the development of innovations that can create economic and social impacts for scaling up at the global level.

As for limitations, this research deals with the SFs of the Android ecosystem. Furthermore, our perspective of analysis is that of qualitative research. Thus, to explore and validate the generalizability of the results, additional case studies would be needed, applying the same analysis framework as used in this paper. Such case studies on, e.g., the iOS ecosystem or the Windows Phone (now called Windows 10 Mobile) ecosystem would be suitable for further research, although the latter has all but lost its relevance. Another option could be to conduct case studies on emergent mobile ecosystems, attempting to identify SFs that could contribute to the rise of new major mobile ecosystems, potentially disrupting the current duopoly of Android and iOS.

Additionally, further research could validate our model through quantitative analysis of data originating from firms in the Android ecosystem, as the measurement and analysis of ecosystem health (*metric **E1.2*** in Table 1), in particular, would benefit from quantitative data, as argued by Den Hartigh *et al.* (2006). A potential new perspective to the research would be to study the social impact of scalable business platforms in the daily life of people.

References

AppBrain (2015). Number of Android applications. Referenced on Jan 19, 2015, http://www.appbrain.com/stats/number-of-android-apps.

Armstrong, M. (2006). Competition in two-sided markets. *RAND Journal of Economics*, **37**(3):668–691.

Armstrong, M. and Wright, J. (2007). Two-sided markets, competitive bottlenecks and exclusive contracts. *Economic Theory*, **32**(2): 353–380.

Arthur, W. (1989). Competing technologies, increasing returns, and lock-in by historical events. *The Economic Journal*, **99**(394):116–131.

Battisti, S. (2012). Social innovation: The process development of knowledge-intensive companies. *International Journal of Services Technology and Management*, **18**(3/4):224–244.

Battisti, S. (2014). Social innovation in living labs: The micro-level process model of public-private partnerships. *International Journal of Innovation and Regional Development*, **5**(4/5):328–348.

Battisti, S. (2019). Digital Social Entrepreneurs as Bridges in Public–Private Partnerships. *Journal of Social Entrepreneurship*, doi: 10.1080/19420676.2018.1541006.

BBC News (2014). Referenced on Sep 30, 2014, http://www.bbc.com/news/technology-29203249.

Bosch, J. (2009). From software product lines to software ecosystems. In *Proceedings of the 13th International Conference on Software Product Lines (SPLC'09)*, San Francisco, CA, USA, 24–28 August 2009, pp. 111–119.

Boudreau, K. (2008). Too many complementors? HEC Paris / London Business School, Working Paper.

Bouwman, H., De Reuver, M., Hampe, F., Carlsson, C. and Walden, P. (2014). Mobile R&D prototypes — What is hampering market implementation? *International Journal of Innovation and Technology Management*, **11**(01), 1440003.

Campbell, P. and Ahmed, F. (2011). An assessment of mobile OS-centric ecosystems. *Journal of Theoretical and Applied Electronic Commerce Research*, **6**(2):50–62.

Canalys (2014). Android on 80% of smart phones shipped in 2013. Referenced on Dec 17, 2014, http://www.canalys.com/newsroom/android-80-smart-phones-shipped-2013.

Cusumano, M. (2010). Technology strategy and management: The evolution of platform thinking: How platform adoption can be an important determinant of product and technological success. *Communications of the ACM*, **53**(1):32–34.

DailyTech (2014). Referenced on Sep 3, 2014, http://www.dailytech.com/Report+Google+to+Force+OEMs+to+Provide+Recent+Android+Builds+or+Lose+Access+to+Google+Apps/article34311.htm.

David, P. (1985). Clio and the economics of QWERTY. *American Economic Review*, **75**(2):332–337.

Davis, J. P. (2016). The group dynamics of interorganizational relationships collaborating with multiple partners in innovation ecosystems. *Administrative Science Quarterly*, **61**(4), 621–661. doi:10.1177/0001839216649350.

Den Hartigh, E., Tol, M. and Visscher, W. (2006). The health measurement of a business ecosystem. In *Proceedings of the European Network on Chaos and Complexity Research and Management (ECCON 2006) Annual Meeting*, Bergen aan Zee, The Netherlands, 20–21 October 2006.

Dyer, J. and Singh, H. (1998). The relational view: Cooperative strategy and sources of interorganizational competitive advantage. *Academy of Management Review*, **23**(4):660–679.

Eisenmann, T. (2008). Managing proprietary and shared platforms. *California Management Review*, **50**(4):31–53.

Eisenmann, T., Parker, G. and Van Alstyne, M. (2006). Strategies for two-sided markets. *Harvard Business Review*, **84**(10):92–101.

Eisenmann, T., Parker, G. and Van Alstyne, M. (2008). Opening platforms: How, when and why? Harvard Business School, Working Paper 09-30.

Forbes (2013). Referenced on Aug 25, 2014, http://www.forbes.com/powerful-brands/list.

Fortune (2014). With 15.5% of smartphone sales in Q1, Apple took 65% of profits. Referenced on Jan 21, 2015, http://fortune.com/2014/05/09/with-15-5-of-smartphone-sales-in-q1-apple-took-65-of-profits.

Gartner (2014). Gartner says worldwide traditional PC, tablet, ultramobile and mobile phone shipments on pace to grow 7.6 percent in 2014. Referenced on Jan 13, 2015, http://www.gartner.com/newsroom/id/2645115.

Gawer, A. (2009). Platform dynamics and strategies: From products to services. *Platforms, Markets and Innovation*, Ed. A. Gawer. Edward Elgar Publishing, pp. 45–76.

Gawer, A. (2014). Bridging differing perspectives on technological platforms: Toward an integrative framework. *Research Policy*, **43**:1239–1249.

Gawer, A. and Cusumano, M. (2008). How companies become platform leaders. *MIT Sloan Management Review*, **49**(2):28–35.

Gawer, A. and Cusumano, M. (2014), Industry platforms and ecosystem innovation. *Journal of Product Innovation Management*, **31**:417–433.

Gawer, A. and Henderson, R. (2007). Platform owner entry and innovation in complementary markets: Evidence from Intel. *Journal of Economics & Management Strategy*, **16**(1):1–34.

Google (2014). Referenced on Mar 4, 2014, http://investor.google.com/pdf/20131231_google_10K.pdf.

Google (2016). Referenced on Dec 21, 2016, http://japan.googleblog.com/2016/07/android-one.html.

Hagiu, A. (2006). Pricing and commitment by two-sided platforms. *RAND Journal of Economics*, **37**(3):720–737.

Hagiu, A. (2009a). Multi-sided platforms: From microfoundations to design and expansion strategies. Harvard Business School, Working Paper.

Hagiu, A. (2009b). Two-sided platforms: Product variety and pricing structures. *Journal of Economics and Management Strategy,* **18**(4):1011–1043.

Hagiu, A. and Spulber, D. (2013). First-party content and coordination in two-sided markets. *Management Science,* **59**(4):933–949.

Heaton, J. (2008). Secondary analysis of qualitative data: An overview. *Historical Social Research/Historische Sozialforschung,* **33**(3), 33–45.

Hyrynsalmi, S., Mäkilä, T., Järvi, A., Suominen, A., Seppänen, M. and Knuutila, T. (2012). App Store, Marketplace, Play! An analysis of multi-homing in mobile software ecosystems. In *Proceedings of the International Workshop on Software Ecosystems (IWSECO 2012),* Boston, MA, USA, 18 June 2012.

Iansiti, M. and Levien, R. (2004). *The Keystone Advantage: What the New Dynamics of Business Ecosystems Mean for Strategy, Innovation, and Sustainability.* Harvard Business School Press, Boston, MA, USA.

Idu, A., van de Zande, T. and Jansen, S. (2011). Multi-homing in the Apple ecosystem: Why and how developers target multiple Apple app stores. In *Proceedings of the International Conference on Management of Emergent Digital EcoSystems (MEDES'11),* San Francisco, CA, USA, 21–23 November 2011.

InfoWorld (2014). Referenced on Jun 17, 2014, http://www.infoworld.com/d/mobile-technology/meet-aosp-the-other-android-while-you-stillcan-236070.

Lee, R. (2013). Vertical integration and exclusivity in platform and two-sided markets. *American Economic Review,* **103**(7):2960–3000.

Luukkainen, S., Karjalainen, M., Winter, J. and Bagheri Majdabadi, M. (2016). Business model analysis of mobile traffic safety services. *INFO,* **18**(1), 56–66.

Magnusson, P. R., Wästlund, E. and Netz, J. (2016). Exploring users' appropriateness as a proxy for experts when screening new product/service ideas. *Journal of Product Innovation Management,* **33**(1), 4–18.

Muegge, S. (2013). Platforms, communities, and business ecosystems: Lessons learned about technology entrepreneurship in an interconnected world. *Technology Innovation Management Review*, **3**:5–15.

Müller, R. M., Kijl, B. and Martens, J. K. (2011). A comparison of inter-organizational business models of mobile app stores: There is more than open vs. closed. *Journal of Theoretical and Applied Electronic Commerce Research*, **6**(2), 63–76.

Parker, G. and Van Alstyne, M. (2005). Two-sided network effects: A theory of information product design. *Management Science*, **51**(10):1494–1504.

Pon, B., Seppälä, T. and Kenney, M. (2014). Android and the demise of operating system-based power: Firm strategy and platform control in the post-PC world. *Telecommunications Policy*, **38**(11), 979–991.

Remneland-Wikhamn, B., Ljungberg, J., Bergquist, M. and Kuschel, J. (2011). Open innovation, generativity and the supplier as peer: The case of iPhone and Android. *International Journal of Innovation Management,* **15**(01): 205–230.

Rochet, J.-C. and Tirole, J. (2003). Platform competition in two-sided markets. *Journal of the European Economic Association*, **1**(4):990–1029.

Rogers, E. M. (2003). *Diffusion of Innovations*, 5th edn. Free Press, NY.

Rumelt, R., Schendel, D. and Teece, D. (1991). Strategic management and economics. *Strategic Management Journal,* **12**:5–29.

Rysman, M. (2004). Competition between networks: A study of the market for Yellow Pages. *Review of Economic Studies*, **71**(2): 483–512.

Rysman, M. (2009). The economics of two-sided markets. *The Journal of Economic Perspectives*, **23**(3):125–143.

SC Magazine (2012). Referenced on Sep 3, 2014, http://www.scmagazine.com.au/News/289242,google-employs-bouncer-to-cleanse-android-malware.aspx.

Stabell, C. and Fjeldstad, Ø. (1998). Configuring value for competitive advantage: On chains, shops, and networks. *Strategic Management Journal*, **19**:413–437.

Sustainable Brands (2013). Referenced on Aug 25, 2014, http://www.sustainablebrands.com/news_and_views/communications/disney-yahoo-googletop-list-100-companies-most-loved-consumers.

Thomas, L. and Autio, E. (2012). Modeling the ecosystem: A meta-synthesis of ecosystem and related. In *Proceedings of the DRUID*

Society Conference 2012, CBS, Copenhagen, Denmark, 19–21 June 2012.

Thomas, L. and Autio, E. (2014). The fifth facet: The ecosystem as an organizational field. In *Proceedings of the DRUID Society Conference 2014*, CBS, Copenhagen, Denmark, 16–18 June 2014.

Thomas, L., Autio, E. and Gann, D. (2014). Architectural leverage: Putting platforms in context. *The Academy of Management Perspectives*, **28**(2):198–219.

Tuunainen, V., Tuunanen, T. and Piispanen, J. (2011). Mobile service platforms: Comparing Nokia OVI and Apple App Store with the IISI^{n} model. In *Proceedings of the 10th International Conference on Mobile Business*, Como, Italy, 20–21 June 2011, pp. 74–83.

Williamson, P. and De Meyer, A. (2012). Ecosystem advantage: How to successfully harness the power of partnerships. *California Management Review*, **55**(1):24–46.

Yin, R. (2003). *Case Study Research: Design and Methods*. Sage Publications, Thousand Oaks, CA, USA.

9to5Google (2011). Referenced on Dec 20, 2016, http://9to5google.com/2011/09/07/shocker-android-has-been-developing-a-lead-device-for-two-years.

Chapter 2

Major Constructs: Culture and Technology Acceptance

Chia-Liang Hung and Jerome Chih-Lung Chou

1. Introduction

Despite continuous growth in the sales of mobile phones and networks worldwide, the growth of mobile commerce applications is slower than expected. The global mobile pioneer Japan NTT DoCoMo is one exception to this trend. From February 1999 to December 2005, 45 million users subscribed to its i-mode mobile commerce services. This figure represents nearly 80% of DoCoMo's mobile users at that time [Natsuno, 2003]. As DoCoMo exported i-mode to Europe, it developed an image similar to that of other successful Japanese products, inevitably becoming a target of scholarly research [Hung and Yeh, 2007]. However, the global spread of i-mode did not match its frantic pace in Japan's domestic market; only 100,000 users had subscribed after the first six months, less than one-fourth of the projected 500,000 [Europemedia 2002]. Some researchers have articulated several obstacles that hinder the speed of overseas i-mode diffusion. Gawer and Cusumano [2002] argued that NTT DoCoMo's local i-mode standard was not adequate as a worldwide standard due to its shortcomings in compatibility and potential for expansions. Funk [2001] and Beck and Wade [2003] specified that the preferences and mindset of i-mode's young target consumers, i.e., high school and college students, facilitated i-mode's diffusion in the Japan. However, this did not prove to be the case in Europe. Further, the cultural infrastructures in different countries influence the consumption values and social communications of consumers confronting new products or services [Blackwell *et al.*, 1995].

For these reasons, mobile innovators and pioneers must study consumer mindset differences, and propose suitable product designs and marketing strategies to increase the adoption rates of mobile commerce, especially when going global. Steinbock [2003] argued that after telecom liberalization, the globalization strategy is becoming increasingly important to mobile operators. This strategy allows companies to achieve economies of scale in investment and to leverage economies of scope in telecommunication experience. For example, Vodafone, Sonera, Telia, Orange, T-mobile, and NTT DoCoMo have all expanded overseas. This is especially helpful for small-country mobile operators such as Finland's Sonera and Sweden's Telia, who, respectively, obtained 40% and 14% of their revenue from overseas markets in 2003 [Steinbock, 2003]. Moreover, the largest mobile operator, Vodafone Group, gained 67% of its revenues overseas in 2000 [Steinbock, 2003]. Facing a continuously decreasing average revenue per user (ARPU), Taiwan's Chunghwa Telecom, a small-country mobile operator, has recently planned to target Vietnam, Thailand, and Cambodia for launching mobile services overseas [Reuters, 2008; CENS, 2009; Telecomsinsight, 2009]. In addition, Southeast Asia should be an investor's paradise. It offers a market of 560 million people, rich natural resources, skilled labor, and an export industry concentrated in global high-growth sectors — all tied together in a free-trade area created by the Association of Southeast Asian Nations (ASEAN), where the value of consumer market equals that of China's booming coastal region in value and is bigger than any other market in Asia [Schwarz and Villinger, 2004]. Since 1994, Taiwan's government has sought to discourage Taiwanese from investing in mainland China through its "Go South" policy, which is a policy to form economic partnerships with neighboring countries in the Southeast Asia [Nanto and Chanlett-Avery, 2006]. Nevertheless, as illustrated by NTT DoCoMo's "global service strategy" [Hung and Yeh, 2007], local institutional diversities can make it difficult to enter international global markets.

Brady [2004] argued that anticipated growth was dependent on the consumer and business adoption, which will be based on the perceived value influenced by contextual factors. Cultural differences necessitate having this flexible orientation to effectively develop and implement global marketing strategies [Harvey *et al.*, 2008]. Yalcinkaya [2008] suggested that social interactions based on culture significantly influence the adoption of new products in every country; the strength of their impact on adoption varies across countries. This paper reviews some of the dominant theoretical frameworks for the adoption of technology such as Davis' technology acceptance model (TAM) [Davis, 1989], Rogers' innovation diffusion theory (IDT) [Rogers, 1995], and the unified theory of acceptance and use of technology (UTAUT) conceptualized by Venkatesh *et al.* [2003], which integrates TAM with IDT. However, these models concentrate on function-driven behavioral intentions, and do not explain consumer choices based on consumer mindset influences on innovation. To use the limited resources of marketing campaigns efficiently, it is critical to choose the best value proposition to appeal to customers' perceived value for a particular cultural circumstance. Hence, the purpose of this chapter is not only to identify how influential cultural orientations enhance the power of TAM, but also to propose cultural marketing strategies for mobile globalization.

2. Dimensions of Culture

Hofstede [1983] argued that attitudes, values, and cognition differ between countries. Adler [1983] also stated that cultural divergences cause lifestyle differences between groups. Hofstede [2001] later developed five cultural orientations based on a large sample of 116,000 employees across more than 60 countries. He found five significant differences in cultures: power distance (PD), uncertainty avoidance (UA), individualism (ID), masculinity (MA), and long-term orientation (LT) [Hofstede, 2001]. Until now, Hofstede's cultural orientations are widely used to compare the national differences for international business. For example, Duque and Lado [2010] use them to compare consumer satisfaction with services across cultures.

The first dimension, PD, represents a culture's degree of acceptance of unequal power distribution or social status. In countries with a larger PD, a few key persons usually govern society in a highly centralized manner. People in such countries often work and live within a hierarchical structure, where one's judgment remains compliant with that of superiors, and managers possess great privilege [Hofstede, 2001].

The second dimension, UA, is the degree of anxiety caused by unclear, risky, or fuzzy circumstances. In a country with higher UA, people are generally more conservative, prefer a safe and stable environment, and conform to public law and social norms [Hofstede, 2001].

The third dimension, ID, emphasizes the pursuit of personal interests, the realization of individual achievements, and the defense of private esteem and independence. For example, the American culture has high ID, while the Japanese culture has low ID, and exhibits more collectivism [Hofstede, 2001].

The fourth dimension, MA, represents an aggressive attitude toward capturing wealth, materialism, growth, and success. People from cultures with high MA prefer a strong backbone to sorrowful distress [Hofstede, 2001].

Finally, the fifth dimension is LT, which emphasizes patience and tolerance. This dimension usually influences people to scrimp within their household, to have patience with current misfortune, to save money for the future, and to respect conventions. In general, the cultures influenced by Confucius in Southeast Asia have a higher tendency toward LT than America: people in Confucian cultures tend to work hard, live frugally, and act prudently rather than strive for immediate results [Hofstede and Bond, 1988].

3. Technology Acceptance Model

The TAM is a theory that models how users accept and use a technology. This model suggests that when users encounter a new technology, a number of factors influence their decision about how and when they will use it. These factors include perceived usefulness (PU) and perceived ease-of-use (PEOU). Davis [1989] defined PU as the degree to which a person believes that using a particular system will enhance his or her job performance, while PEOU is the degree to which a person believes that using a particular system will be effortless. Figure 1 illustrates the basic framework of the TAM.

TAM2 extends the original TAM to identify the causes of PU and intentions to use in terms of social influence and cognitive instrumental processes [Venkatesh and Davis, 2000]. The former includes subjective norms, voluntariness, and image, while the latter covers job relevance, output quality, and result demonstrability. The UTAUT model further integrates several competing user acceptance models [Venkatesh *et al.*, 2003]. Based on TAM2, this model adds moderators for sex, age, experience, and voluntariness, and thus outperforms each of its constituent models. The UTAUT model achieves a higher adjusted R square

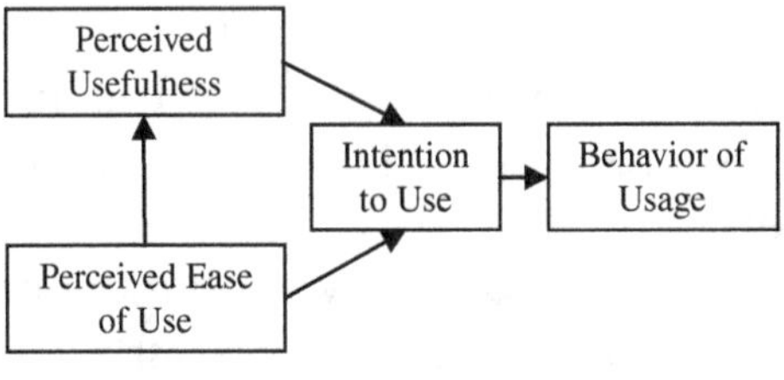

Figure 1. Technology acceptance model.
Source: Davis [1989].

of 69% of variance of intentions to use. Srite and Karahanna [2006] used Hofstede's culture orientations as UTAUT moderators to study the factors influencing the adoption of PCs and PDAs, and discovered some cultural influences on subjective norms and PU. Thus, including external factors that influence the user's perceived value of innovation, such as social factors and the espoused culture, in the TAM may enable a better explanation of how users accept and use technology. Mafé *et al.* [2010] also apply TAM with the theory of planned behavior to compare the adoption rate of mobile commerce (e.g., SMS) among different countries, and consequently find significantly different influential patterns.

4. Relationship between National Culture and TAM

This study defines mobile commerce as business operations conducted by way of mobile devices such as mobile phones, PDAs, etc., over which wireless voice and data are transmitted [Heijden and Valiente, 2002; Skiba *et al.*, 2000]. Thus, the research target focuses on the cultural influences on the innovation adoption of mobile technology. Technology that is most relevant to an organization's demands and feasible for the adopter's smooth operation provides a greater incentive of adoption [De Man and van den Toorn, 2002]. People in a centralized society with a higher PD generally have a lower desire to envisage organizational objectives and societal demands or persuade themselves to challenge existing behavior [De Man and van den Toorn, 2002]. As a result, a higher PD negatively influences PU and PEOU for innovation, and thus decreases the innovation adoption intentions [Waarts and van Everdingen, 2005]. Accordingly, the cultural orientation of PD may positively or negatively influence user adoption intentions. However, the mobile commerce innovation is a bottom-up application especially adopted by young users with a high degree of self-discretion but less relevance to organizational or national demands [Funk, 2001; Beck and Wade, 2003]. The external atmosphere under a centralized society may inhibit users to challenge the hierarchy and prevent them from adopting innovations personally too early. Hence, for the case of mobile commerce, this study proposes the following hypotheses:

H1a: The cultural orientation of PD may negatively influence the PU of mobile commerce innovation on the intention to use of mobile commerce.

H1b: The cultural orientation of PD may negatively influence the PEOU of mobile commerce innovation on the intention to use of mobile commerce.

The degree of UA is usually negatively related to innovation adoption willingness [De Man and van den Toorn, 2002]. The unwillingness of adoption arises from the fear, uncertainty, and doubt (FUD) factor regarding the needs or problems the new technology will address, and how well it will meet those needs [Moore, 2002]. Hence, the cultural orientation of UA often limits the PU of technology innovation. Research on internet shopping also reveals that higher UA increases the fear of trying online purchasing, and thus generates a lower level of PEOU [Zalka, 1999]. Hence, for the case of mobile commerce, this study proposes the following hypotheses:

H2a: The cultural orientation of UA negatively influences the PU of mobile commerce innovation on the intention to use of mobile commerce.

H2b: The cultural orientation of UA negatively influences the PEOU of mobile commerce innovation on the intention to use of mobile commerce.

The ID orientation is positively related to personal intentions to pursue innovative technology for problem solving, convenience, and performance enhancement [Sørnes *et al.*, 2004]. Individualistic cultures usually encourage people to emphasize efficiency, speed, and performance using the most useful and adaptable means [Hofstede, 2001]. Hence, for the case of mobile commerce, this study proposes the following hypotheses:

H3a: The cultural orientation of ID positively influences the PU of mobile commerce innovation on the intention to use of mobile commerce.

H3b: The cultural orientation of ID positively influences the PEOU of mobile commerce innovation on the intention to use of mobile commerce.

People with high MA prefer to achieve objectives, and aggressively pursue promotion and wealth [Hofstede and Bond, 1988]. Therefore, the criteria of efficiency, convenience, and identity are important decision factors. Research on the advertising effects caused by differences in national cultures shows that countries with higher MA reflect a greater emphasis on advertisements appealing to convenience and efficiency [Albers-Miller and Gelb, 1996]. People who like shopping online are more likely to have the characteristic of MA and more willing to try novel tools [Zalka, 1999]. Hence, for the case of mobile commerce, this study proposes the following hypotheses:

H4a: The cultural orientation of MA positively influences the PU of mobile commerce innovation on the intention to use of mobile commerce.

H4b: The cultural orientation of MA positively influences the PEOU of mobile commerce innovation on the intention to use of mobile commerce.

The LT orientation emphasizes patience and tolerance [Hofstede, 2001]. An organizational attitude of enduring current misfortune or discomfort to invest in the future leads the organization to confront the challenges of innovation and learn novel technologies despite difficulties [Waarts and van

Everdingen, 2005]. Hence, for the case of mobile commerce, this study proposes the following hypotheses:

H5a: The cultural orientation LT positively influences the PU of mobile commerce innovation on the intention to use of mobile commerce.

H5b: The cultural orientation LT positively influences the PEOU of mobile commerce innovation on the intention to use of mobile commerce.

This study extends the original TAM by adding Hofstede's five cultural orientations as moderators. However, as in prior research [Srite and Karahanna, 2006], this study only focuses on adoption intentions to determine the social influences on PU and PEOU, as Figure 2 illustrates.

This study uses several question items corresponding to each construct in Figure 2 to determine the cultural moderation on the TAM in the case of mobile commerce. These items are based on the original Davis measurements and modified by some TAM-related studies [Chau, 1996; Davis, 1989; Mao and Palvia, 2006; Venkatesh and Davis, 2000]. Five items measure PU: ease of perception, productivity, efficiency, performance, and task facilitation. Another five items measure PEOU: ease of learning, ease of practice, ease of interaction, flexibility, and ease of accomplishment. In addition, measurements of intention to use focus on the anticipation of new trials [Davis, 1989;

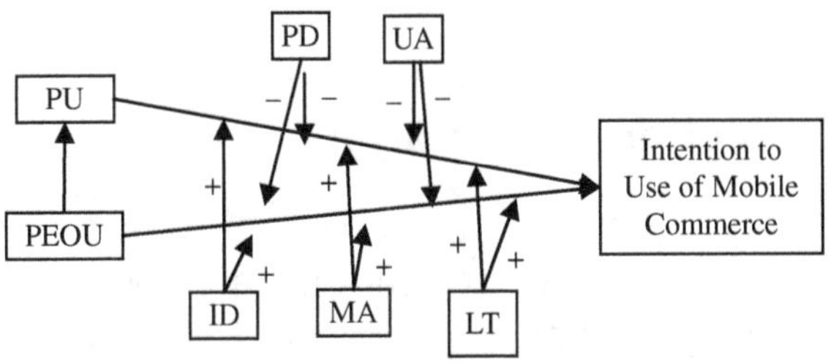

Figure 2. Research model.

Taylor and Todd, 1995; Venkatesh *et al.*, 2003] and increasing usage [Bhattacherjee and Sanford, 2006].

Measurements of PD emphasize the degree of centralization [Hofstede and Bond, 1988], respect for authoritative leadership [Waarts and van Everdingen, 2005], compliance with superior's orders [Srite and Karahanna, 2006], and opinion regarding superior's instructions [Huang, 2003]. Items referring to the arguments of Hofstede and related studies measured UA, concentrating on the conformity to rules and regulations, resistance to change, and response to stress [Adler, 1983; Hofstede and Bond, 1988; Srite and Karahanna, 2006]. The ID orientation focuses on personal achievement, egocentricity, freedom, and risk ventures, while the MA orientation includes social status, job identity, career achievement, and reward [Bouaziz, 2008; Srite and Karahanna, 2006]. For the LT-oriented orientation, measurement items target the characteristics of prudence, persistence, futurity, and adaptation [Adler, 1983; Hofstede and Bond, 1988; Waarts and van Everdingen, 2005]. For a detailed list of all questionnaire items. The interviewees answered each question using an adjusted Likert 7-scale, where 1 is "not at all" and 7 is "very much".

For measuring the moderating effects, this research adopted the Chin *et al.* [2003] approach; cross-products were created by multiplying indicators of each interacting constructs. For example, the product indicator (i.e., PD $\times$ PU) was developed by creating the product of standardized scores of the predictor PD and the moderator variable PU.

5. Data Collection and Analysis

A questionnaire webpage was used to collect interviewees' responses. To obtain a larger sample and allow subsequent cross-cultural comparisons, the investigators posted invitations on several famous Bulletin Board Stations (BBSs) and community websites in Taiwan and Malaysia. Malaysia is a representative country of high economic growth with great opportunities in the area of Southeast Asia beyond four Asian Tigers, which are Korea, Taiwan, Hong Kong, and Singapore. However, according to the study of Hofstede [2001], the cultural dimensions between Taiwan and Malaysia were not similar in 1970s. Malaysia performed higher scores of PD, ID, and MA, while Taiwan had a higher score of UA. Thus, this study reinvestigates the current Malaysian cultural data, especially, the young group of targeting segment of mobile commerce. The Taiwan sample mainly comes from PTT (www.ptt.cc), the biggest BBS community, WeFong (bbs.wefong.com), a famous forum for mobile phones, Gamebase (www.gamebase.com.tw), a hot mobile software website, and Youthwant (www.youthwant.com.tw), a large community for youth, and especially college students. The Malaysian sample includes the Sabah Forum (www.e-sabah.com), CARI (mforum5.cari.com.my), and Webportal Malaysia (www.webportal.com.my). The data collection process spanned from April to June of 2009. The convenience survey approach in this study collected 479 samples, including 265 Taiwanese users and 214 Malaysian users.

The survey results are consistent to the prior research which argued that the young group is the mobile target [Funk, 2001; Beck and Wade, 2003]. Moreover, most respondents went surfing the Internet on mobile devices for an average of two hours per day. Taiwan users preferred mobile information searching, while Malaysia users preferred mobile email services.

6. Findings and Discussion

This study includes a multiple regression analysis of three control variables, age, gender, and experience, two explanatory variables, PU and PEOU, and one performance variable, intention to use, in Model 1. Model 2 adds 10 product indicators: PD × PU, PD × PEOU, UA × PU, UA × PEOU, ID × PU, ID × PEOU, MA × PU, MA × PEOU, LT × PU, and LT × PEOU, to the original Model 1 for examining five cultural moderation effects on PU and PEOU, respectively. Both models confirm the positive relation from PU to IU and the positive one from PEOU to PU that had been discovered by many previous studies. However, the *R*-square coefficient of Model 2 explaining the influences from PU and PEOU to IU is significantly higher than that in Model 1, because the former includes five cultural orientations. This shows that the cultural orientations enhanced the explanatory power of the usage intentions of mobile commerce. The variable of PU is still a significant factor in the usage intentions. However, some regression coefficients reveal different cultural moderation effects on the PU and PEOU variables between Taiwan and Malaysia. For Taiwanese users, the cultural orientations of ID, MA, and LT positively modulate the PU variable with significant coefficients of 0.207, 0.298, and 0.283, respectively, while PD and UA have a negative influence with significant coefficients of –0.283 and –0.277, respectively. Moreover, PD, ID, and LT positively modulate the PEOU with significant coefficients of 0.281, 0.246, and 0.285, respectively, while UA and MA have a negative influence with significant coefficients of –0.247 and –0.375, respectively. For Malaysian users, the cultural orientations of PD, ID, and LT positively influence the PU with significant coefficients of 0.223, 0.424, and 0.342, respectively, while only UA has a negative with a significant coefficient of –0.346. Moreover, ID and LT positively influence the PEOU with significant coefficients of 0.479 and 0.438, respectively, while PD and UA have a negative influence with significant coefficients of –0.327 and –0.405, respectively. However, MA does not have a significant effect on either PU or PEOU for Malaysian users.

Results generally support hypotheses H2a, H2b, H3a, H3b, H5a, and H5b. In other words, the UA orientation mitigates the effects of PU and PEOU on the intention to use mobile commerce, while ID and LT enhance the effects of PU and PEOU on the intention to use mobile commerce. However, MA only had a positive influence on PU for Taiwanese users, partially supporting H4a. Moreover, MA had a significant negative influence on PEOU for Taiwanese users, rejecting H4b. The most interesting results involve the influences of PD on PU and PEOU. For the Taiwanese, PD has a negative effect on PU but a positive effect on PEOU. However, Malaysians have a positive effect of PD on PU but a negative effect on PEOU. Countries with a higher PD may promote the adoption toward innovation [Harris *et al.*, 2005]. People in this type of society may be more willing to pursue organizational objectives, including the adoption of new technology. Kumar *et al.* [2007] showed that subordinates within a hierarchy of higher PD are more willing to conform to social norms or the orders of superiors. Neither Taiwan nor Malaysian data fully support the arguments of Hypotheses H1a and H1b, which claim that PD has either a positive influence on both PU and PEOU or a negative influence on PU and PEOU. These results partially support hypotheses H1a and H1b. As a result, the exact moderation effects of PD on PU and PEOU remain unclear.

7. Conclusions

For innovations related to collective actions and underlying mindsets, the culture-embedded TAM is capable of capturing multiple facets. This study expands the TAM by including cultural factors even though the results do not support some of the moderation hypotheses. Nonetheless, the results provide strategic implications for globalized mobile operators. Steinbock [2003] argued that after telecom liberalization, the globalization strategy is becoming increasingly important to mobile operators wanting to achieve economies of scale and economies of scope in investments. For example, Vodafone, Sonera, T-mobile, and NTT DoCoMo all actively operate in oversea markets. Nevertheless, local institutional diversities may make global penetration difficult [Beck and Wade, 2003; Gawer and Cusumano, 2002]. For this reason, the cultural orientations on the adoption of mobile commerce identified in this study using empirical data from Taiwan and Malaysia will be helpful to mobile operators entering international markets. The results highlight the cultural influence of ID on innovation adoption, especially for the young. This has become the prevailing custom in the global wave of telecom liberation, even in Eastern countries. Hence, mobile marketing that accentuates egocentricity and freedom may capture the attention of young users and encourage them to adopt mobile commerce. Both this study and Hofstede's [2001] study show most Asian countries have the characteristics of UA. Therefore, Western mobile operators planning to touch Eastern markets must initiate marketing programs of orientation, education, and assurance to reduce users' perceived risks. Moreover, the LT in most Asian countries molded by the Confucius culture drives people to scrimp now but invest in the future. Thus, mobile commerce marketing campaigns should focus on the attraction of futurity. Moreover, since most Taiwanese pursue the characteristics of MA, mobile operators should launch mobile diffusion campaigns that emphasize the benefits of career achievements and capturing wealth. However, this study shows that the persuasion power of MA may not be as effective in enhancing the adoption of mobile commerce in Malaysia. As the empirical testing of this study shows opposite moderation influences of PD on the adoption of mobile commerce to Taiwanese and Malaysian. A transnational mobile operator should not manipulate the same value position based on the PD orientation for the sake of increasing the possibility of mobile adoption. Taiwan is a long-term export-oriented economic entity with global commerce relationships with Western countries, especially North America. As a result, it has transformed the traditional Chinese bureaucratic structure into a flat hierarchic and democratic society with the direct voting right in presidential elections. The adoption behavior may be similar to the arguments of Waarts and van Everdingen [2005]: the less the PD of a society is, the more the intentions to use innovation will be. Therefore, the value proposition of mobile commerce using the rise of bourgeoisie seems to be persuasive to Taiwanese. However, this promotional method may not be effective for Malaysian users, who exhibit more compliance with superiors, and accept authoritative leadership. Accordingly, a transnational mobile operator entering this kind of area should launch marketing campaigns in partnership with governmental technology-diffusion championship program. This is a type of the coercive top-down mission suggested by Kumar *et al.* [2007]. In summary, considering the potential cultural influences on innovation adoption can not only reduce local conflicts, but also make mobile diffusion strategies more effective.

References

Adler, N. J. (1983). Cross-cultural management research: The ostrich and the trend. *Academy of Management Review*, **8**(2):226–232.

Albers-Miller, N. D. and Gelb, B. D. (1996). Business advertising appeals as a mirror of cultural dimensions: A study of eleven countries. *Journal of Advertising*, **25**(4):57–70.

Beck, J. C. and Wade, M. (2003). *DoCoMo — Japan's Wireless Tsunami: How One Mobile Telecom Created a New Market and Became a Global Force*. AMACOM, NY.

Bhattacherjee, A. and Sanford, C. (2006). Influence processes for information technology acceptance: An elaboration likelihood model. *MIS Quarterly*, **30**(4):805–825.

Blackwell, R. D., Engel, J. F. and Miniard, P. W. (1995). *Consumer Behavior*. South-Western, Mason, OH.

Bouaziz, F. (2008). Public administration presence on the web. *The Electronic Journal of e-Government*, **6**(1):11–22.

Brady, M. (2004). Book review: mobile commerce: technology, theory and application. *International Journal of Service Industry Management*, **15**(2):220–224.

CENS (2009). Chunghwa Telecom plans to step into Southeast Asian market. Referenced on Jun 20, 2010, http://cens.com/cens/html/en/news/news_inner_30317.html.

Chin, W., Marcolin, B. and Newsted, P. (2003). A partial least squares latent variable modeling approach for measuring interaction effects: Results from a Monte Carlo simulation study and an electronic-mail emotion/adoption study. *Information Systems Research*, **14**(2):189–217.

Chau, P. Y. K. (1996). An empirical assessment of a modified technology acceptance model. *Journal of Management Information System*, **13**(2):185–204.

Davis, F. D. (1989). Perceived usefulness, perceived ease of use, and user acceptance of information technology. *MIS Quarterly*, **13**(3): 318–340.

De Man, W. H. E. and van den Toorn, W. H. (2002). Culture and the adoption and use of GIS within organizations. *International Journal of Applied Earth Observation and Geoinformation*, **4**(1): 51–63.

Duque, L. C. and Lado, N. (2010). Cross-cultural comparisons of consumer satisfaction ratings: A perspective from Albert Hirschman's theory. *International Marketing Review*, **27**(6):676–693.

Europemedia (2002). Take up of KPN mobile's i-mode slower than expected. Aug. 13, Amsterdam.

Funk, J. L. (2001). *The Mobile Internet: How Japan Dialed Up and the West Disconnected*. AMACOM, ISI Publications, NY.

Gawer, A. and Cusumano, M. A. (2002). *Platform Leadership: How Intel, Microsoft, and Cisco Drive Industry Innovation*. Harvard Business School Press, Boston, MA.

Harris, P., Rettie, R. and Kwan, C. C. (2005). Adoption and usage of M-commerce: A cross-cultural comparison of Hong Kong and The United Kingdom. *Journal of Electronic Commerce Research*, **6**(3): 210–224.

Harvey, M. G., Kiessling, T. S. and Richey, R. G. (2008). Global social time perspectives in marketing: A strategic reference point theory application. *International Marketing Review*, **25**(2):146–165.

Heijden, H. V. d. and Valiente, P. (2002). The value of mobility for business process performance: Evidence from Sweden and the Netherlands. *European Conference on Information Systems (ECIS2002)*, Poland, 6–8 June 2002, pp. 1144–1153.

Hofstede, G. (1983). National cultures in four dimensions: A research-based theory of cultural differences among nations. *International Studies of Management & Organization*, **13**(1/2):46–75.

Hofstede, G. (2001). *Culture's Consequences: Comparing Values, Behaviors, Institutions, and Organizations across Nations*. Sage Publications, Thousand Oaks, CA.

Hofstede, G. and Bond, M. H. (1988). The Confucius connection: From cultural roots to economic growth. *Organization Dynamics*, **16**(4): 4–21.

Huang, L. (2003). The impact of cultural values on E-mail acceptance: Evidence from PRC. Unpublished Doctoral Dissertation, Lingnan University, Hong Kong.

Hung, C. L. and Yeh, S. Y. (2007). The diffusion of i-mode overseas: A perspective of institutional differences. *International Journal of Innovation and Technology Management*, **4**(1):69–86.

Kumar, V. Mukerji, B. Butt, I. and Persaud, A. (2007). Factors for successful e-Government adoption: A conceptual framework. *Electronic Journal of e-Government*, **5**(1):63–76.

Mafé, C. R., Blas, S. S. and Tavera-Mesías, J. F. (2010). A comparative study of mobile messaging services acceptance to participate in television programs. *Journal of Service Management*, **21**(1): 69–102.

Mao, E. and Palvia, P. (2006). Testing an extended model of IT acceptance in the Chinese cultural context. *ACM SIGMIS Database*, **37**(2): 20–32.

Moore, J. (2002). *Crossing the Chasm: Marketing and Selling Technology Products to Mainstream Customers.* Harper Business, NY.

Nanto, D. K. and Chanlett-Avery, E. (2006). The rise of China and its effect on Taiwan, Japan, and South Korea: U.S. policy choices. Referenced on June 20, 2010, http://www.fas.org/sgp/crs/row/RL32882.pdf.

Natsuno, T. (2003). *i-mode Strategy.* John Wiley & Sons, Chichester, UK.

Reuters (2008). Taiwan's Chunghwa Telecom to enter Cambodia. Referenced on Dec 20, 2009, http://www.reuters.com/article/idUSTP33960420080612.

Rogers, E. M. (1995). *Diffusion of Innovations*, 4th edn. Free Press, NY.

Schwarz, A. and Villinger, R. (2004). Integrating Southeast Asia's economies. Referenced on June 20, 2010, http://www.mckinseyquarterly.com/Integrating_Southeast_Asias_economies_1392.

Skiba, B., Johnson, M. and Dillon, M. (2000). *Moving in Mobile Media Mode*, Lehman Brothers, NY. Referenced on December 20, 2009, http://www.estsoftware.com.

Sørnes, J.-O., Stephens, K. K., Sætre, A. S. and Browning, L. D. (2004). The reflexivity between ICTs and business culture: Applying Hofstede's theory to compare Norway and the United States. *Informing Science Journal*, **7**:1–29.

Srite, M., and Karahanna, E. (2006). The role of espoused national cultural values in technology acceptance. *MIS Quarterly*, **30**(3): 679–704.

Steinbock, D. (2003). *Wireless Horizon: Strategy and Competition in the Worldwide Mobile Marketplace*. AMACOM, NY.

Taylor, S. and Todd, P. A. (1995). Understanding information technology usage: A test of competing model. *Information System Research*, **6**(2):144–176.

Telecomsinsight (2009). Chunghwa Telecom sets sights on Vietnam and Thailand. Referenced on December 20, 2009, http://www.telecomsinsight.com/file/84445/chunghwa-telecom-sets-sights-on-vietnam-and-thailand.html.

Venkatesh, V. and Davis, F. D. (2000). A theoretical extension of the technology acceptance model: Four longitudinal field studies. *Management Science*, **46**(2):186–204.

Venkatesh, V., Morris, M. G., Davis, G. B. and Davis, F. D. (2003). User acceptance of information technology: Toward a unified view. *MIS Quarterly*, **27**(3):425–478.

Waarts, E. and van Everdingen, Y. (2005). The influence of national culture on the adoption status of innovations: An empirical study of firms across Europe. *European Management Journal*, **23**(6): 601–610.

Yalcinkaya, G. (2008). A culture-based approach to understanding the adoption and diffusion of new products across countries. *International Marketing Review*, **25**(2):202–214.

Zalka, L. M. (1999). *Globalization of the Internet: Convergence or a multicultural community?* Unpublished Doctoral Dissertation, Florida International University, Florida.

Chapter 3

Major Constructs: Brands

Shwu-Ing Wu and Li-Pang Ho

1. Introduction

Since the rapid changes in technology, the product life cycle has become shorter, that makes enterprises face increasingly complex and unpredictable competitive environment. At this moment, an enterprise will lose market opportunities or face the dilemma of sales recession if it has no capacity of innovation. Therefore, if an enterprise wants to stand out on the competitive market, it must continuously strive for innovation. Peter Drucker has said "innovate or die", meaning that innovation is the main source of competitive advantage [Prajogo and Sohal, 2006]. Shelton [2009] pointed out that many leaders of well-known brands approved that innovative products and services can generate greater customer value, increase customer's brand preference, and create more sales opportunities. However, "innovation" is not just a corporate strategy, it should be evaluated based on consumer's viewpoint. If the innovations of product or service cannot perceived by consumer, it is difficult to succeed.

In addition to consumer perception toward innovative products, the brand awareness is another important factor on purchase intention of innovative products. Because the brand awareness will affect the consumer's cognition and attitude toward a new product, a famous brand often provides higher sense of trust that can bring positive consumer perception to complete the final deal [Aaker, 1996]. Lane [2000] confirmed that brand awareness has significant benefit on the new product development; it indicated that brand awareness is the one of most important decision-making factors on purchase intention toward an innovative product.

2. Background

2.1. *Perceived innovation*

Innovation is a process that includes innovating, employing new and useful things, knowledge, or relevant information [Schumpeter, 1934]. Consumer's perception on product innovation can be divided into two viewpoints, producer perspective, and the consumer perspective. On the producer perspective, "innovation" indicates the concept creating for novel products or procedures, so innovation is, namely, to introduce new products, new procedures, or new services into the market [Betz, 1993]. This study is based on the consumer perspective of including the consumer's perceptions on new products' characteristics, adoption risks, and the change-necessary behavior [Danneels and Kleinschmidt, 2001].

Veryzer [1998] addressed that innovation was a process to create new products, services, or procedures, and any new method to create values. Johne [1999] classified innovation into three categories: products (new materials and new products), markets (new markets and new applications), and procedures (management procedures and administrative procedures). Johannessen *et al.* [2001] deemed that there were six types of innovation including new products, new services, new methods or production (manufacturing/technological innovation), new market development, new supply sources, and new organizational methods. When investigating innovation-based competitive strategy, Weerawardena [2003] argued that innovation indicated gradual improvement in the products, production, services, organizational system, and marketing system that help customers to create their values. In terms of technical novelty and the novelty toward the market, Brentani [2001] divided the innovation of new products and new services into suspended innovation and gradually progressing innovation. Therefore, innovation does not only narrowly indicate technology. Instead, it includes marketing, products, and the methods for providing services. In addition, any type of innovation influences the growth of an enterprise positively, and it is the critical factor of success [Schepers *et al.*, 1999]. In summary, innovation can improve the quality of products or services to further advance the image of an enterprise and maintain the loyalty of customers. However, enterprises should endeavor to improve the perception of customers for product innovation.

Rogers [1995] defined innovation by consumer perspective, and he addressed that innovation indicated the novel ideas or products presumed by potential consumers. Investigating from the customer perspective, Kwaku [1995] found that product innovation indicated the use experience of customers and the accommodation of their consumption model for a new product, in which the lower the accommodation of customers was, the greater novelty a product had. Furthermore, product innovation may cause customers to have different usage methods or cause consumers to increase perceived risks, which should not be ignored as well [Ali *et al.*, 1995].

In the Diffusion of Innovation Theory, Rogers [1995] argued that innovation had five characteristics, i.e., relative advantage, compatibility, complexity, triability, and observability. Enterprises may investigate the perception of consumers on innovative products according to their feeling about these five characteristics. Monroe [1991] addressed that in terms of the innovation features of information technology, three factors, including relative predominance, compatibility, and complexity, tended to influence consumers more.

This study was based on the consumer perspective, and the innovation perceived by consumers on innovative technological products was regarded as the basis for the measurement of perceived innovation. Consequently, the features of perceived innovation brought up by Rogers [1995] and Moore *et al.* [1991], including relative advantage, compatibility, complexity, triability, and observability, as well as the items brought by Kwaku [1995], Weerawardena [2003], and Johannessen *et al.* [2001], including familiarity, product predominance/uniqueness, product development, marketing, service, and the integrated effects of new products were revised and adjusted in this study to construct four dimensions, respectively, marketing, product, technology, and services, to develop the questions of consumer perception on innovative products and to measure the perceived innovation of consumers.

2.2. *Brand awareness*

Brand awareness indicates the capability of consumers to identify and memorize a brand in a certain product category [Aaker, 1991]. Brand awareness represents the accumulated memory intensity of consumers for a certain brand [Biswas, 1992]. It influences consumers significantly, and consumers may regard it as the basis for prior consideration while selecting a new product [Hoyer and Brown, 1990]. Therefore, brand awareness is critical when consumers purchase new products [Dodds *et al.*, 1991; Grewal *et al.*, 1998].

The importance of brand awareness lies in that consumers include it into their purchase decision and evaluate it, so an unnoted brand is relatively weak in the consideration of consumers [Aaker, 1991], and a famous brand can effectively reduce the perceived risks of consumers and increase their positive evaluation on the products [Shimp and Bearden, 1982; Rao and Monroe, 1989]. A brand with high awareness and a fine image can advance the loyalty of consumers for the brand, increase the trust of consumers in its products, and strengthen the purchase intention [Aaker, 1991]. Berry [2000] proposed that the famous brand could effectively help consumers identify the product and endorse the value.

Keller [1993] pointed out that brand awareness was composed of brand recall and brand recognition, which can be measured by indicators such as: if one heard the brand, if one was familiar with the brand, and if one knew the characteristics of the brand [Aaker, 1996]. Berry [2000] brought up two dimensions for measuring brand awareness, namely, the contact with an external brand and customer experience. This study refers the measurement items proposed by Aaker [1996] and Berry [2000] to design the measuring dimensions and indices of brand awareness such as the level of the awareness of the brand name in this product categories, if this brand was heard, the meaning represented by the brand, and personal opinions on the brand.

2.3. *Perceived quality*

Perceived quality indicates the expectation of consumers for quality when they purchase goods by their own experience and information, and it will further influence the purchase behavior [Bredahl *et al.*, 2004]. Garvin [1983] divided quality into manufacturing quality and perceived quality. Manufacturing quality is based on products and manufacturing, whereas perceived quality evaluates quality from the perspective of consumers. Steenkamp [1990] defined perceived quality as the special judgment generated when consumers consciously or unconsciously dealt with clues relevant to the attributes of quality. Bhuian [1997] deemed that perceived quality as the judgment or evaluation that consumers had for the consistence of the specifications of a product, or the good and bad attached to the functions of the product. Whether a product is evaluated as high quality or low quality depends on the perception and evaluation of consumers on the product.

Aaker [1996] pointed out that there were three reasons for perceived quality to be different from actual quality. First, the previous impression of consumers influenced the judgment on quality. Secondly, there were differences between the perceptions of manufacturers and consumers for quality. Third, due to incomplete information, consumers might not able to make accurate judgment. Zeithaml [1988] argued that product quality could be divided into internal attributes, namely the product itself, and external attributes, such as value and brand. Zeithaml and Bitner [2000] pointed out that customer satisfaction would be influenced by service quality and product quality; perceived quality was the judgment of customers on the overall excellence of products or services.

Since perceived quality will be influenced by the subjective perception of consumers, environmental contexts, product clues, etc., scholars have proposed that the perceived quality of products should be measured from the perspective of consumers [Garvin, 1983; Zeithaml, 1988]. In this study, the measurement items of perceived quality were developed by referring to the four indices, respectively, outstanding quality, credibility, reliability, and quality consistency, which were brought up by Petrick [2002].

2.4. *Perceived value*

Kotler [1994] argued that if an enterprise would like to maintain the leading predominance, it had to create not only new product characteristics but also new perceived value of customers, which indicated that perceived value of customers would be the focus that enterprises cared the most in the future. Perceived value is a critical influential factor in the purchase and utility of products and services of consumers, the most important index of the repurchase intention of consumers, and the critical factor for enterprises to obtain their competitive advantages [Petrick, 2002].

Zeithaml [1988] addressed that perceived value was the evaluation that consumers had for the overall effectiveness of products based on the perception they gained and paid. Hence, perceived value is the balance between "give" and "get" that consumers have for a product or service [Dodds *et al.*, 1991]. The perception of "give" and "get" of consumers varies from person to person, and the subjective perception that consumers have when they purchase products or services is, namely, perceived value, which cannot be measured or explained by a single index [Monroe, 1973; Jacoby and Olson, 1997].

In this study, the subjective perspective of consumers was employed to measure the perception of the value of innovative products, namely, perceived value, which implies the overall feeling of consumers for the difference between the profit they got from and the resources they gave for the innovative products. Moreover, by referring to the items addressed by Petrick [2002], the measurement indices of perceived value were designed, such as one great purchase, this product is valuable, is worth the money, considerably good value, extra value, reasonable value, economic, and a fine transaction.

2.5. *Purchase intention*

Fishbein and Ajzen (1975) deemed that intention was, namely, the subjective ratio for an individual to perform a particular behavior. If this concept is extended and applied to purchase intention, it indicates the ratio that people are willing to perform a particular purchase behavior. Purchase intention generally means the possibility that consumers would like to purchase a product, which can be used to predict the occurrence of actual purchase behavior [Zeithaml, 1988; Dodds *et al.*, 1991; Grewal *et al.*, 1998]. Consumers with positive purchase intention will form positive commitment which will strengthen the desire of consumers to maintain the trade relationship with sellers [Moorman *et al.*, 1992]. The higher the purchase intention is, the greater the purchase ratio will be [Dodds *et al.*, 1991], and purchase intention may be regarded as the most precise item to predict real purchase behavior [Morwitz and Schmittlein, 1992].

Scholars, such as Zeithaml [1988] and Dodds *et al.* [1991] all deemed that the purchase intention of consumers usually depended on the profits and value they perceived. In the process that consumers make their decisions to purchase a product or service, they will first perceive the quality of the product or service and later have the purchase intention emerged; the higher the perceived value that consumers have for the product, the greater the purchase intention to buy the product will be [Krugman, 1965]. Dodds *et al.* [1991] brought up the items for the measurement of purchase intention, including the possibility to purchase the product, being willing to purchase it by this price, purchase ratio, and purchase intention. In this study, they were revised into four items, respectively, "purchase possibility is high", "I am thinking about purchasing it", "I am willing to purchase it by this price", and "I will recommend other people to purchase it" to be the variables of purchase intention of consumers.

3. Research Hypotheses

3.1. *The influence of perceived innovation on the perceived quality and purchase intention*

The needs of consumers unceasingly change, and their demands for quality continuously increase, so the speed to weed through the old products to bring forth the new ones becomes faster, and all enterprises thus hope to attract consumers to purchase their products through various product innovations. Innovation aims to improve the quality of products or services, and it may be regarded as a measure to advance quality [Johannessen *et al.*, 2001]. Consequently, innovation is a fundamental element in order to improve the quality of a product or create brand-new products, services, or functions to enhance the efficiency of utility [Veryzer, 1998; Johne, 1999]. Innovation includes not only manufacturing technologies but also product marketing and the methods for providing services, and innovation for improvement in the quality of products and services is still more the goal that enterprises pursue [Schepers *et al.*, 1999]. In summary, enterprises may improve their quality of products or services by means of innovation to further increase the purchase intention and loyalty of consumers. The effects of product innovation should thus be considered from the perspective of consumers, and it should measure the perceptions of consumers for the quality of new products and their purchase intention of the new products [Danneels and Kleinschmidt, 2001].

Kwaku [1995] discovered that the more novel customers felt a new product was, the more they would feel the quality of the new product was improved. According to the research of Ottenbacher and Gnoth [2005], innovation can improve the quality of products or services, it possesses indirect benefits in advancing corporate image, and it can increase the loyalty of customers and attract the purchase of potential customers. The investigation by Horn and Salvendy [2006] also addressed that if consumers could be provided with more information on innovative products, their purchase intention would be positively stimulated. Thus, we propose the following hypotheses:

H1a: Consumer-perceived innovation has positive impact on perceived quality toward new products.
H1b: Consumer-perceived innovation has positive impact on purchase intention toward new products.

3.2. *The influence of brand awareness on the perceived quality and purchase intention*

Bitta *et al.* [1981] pointed out that a product with high brand awareness could increase the perceived quality of consumers, namely, a product with high brand awareness brought consumers the feeling of high quality. The study of Wall *et al.* [1991] also showed that consumers perceived higher quality for a highly famous brand. Samiee [1994] found that even if consumers faced an unfamiliar product, they might have positive evaluation for the product when it was of high brand awareness. When investigating the consumer behavior of calculators and walkmans, Dodds *et al.* [1991] also discovered that brand awareness had positive influence on the perceived quality of consumers. Taking bicycles as the experimental objects and conducting a questionnaire survey of undergraduates, Grewal *et al.* [1998] found that there was indeed significantly positive correlation between awareness and perceived quality.

In addition, a great number of studies all proved that brand awareness had significant positive influence on the perceived quality and purchase intention of consumers [e.g., Jacoby and Olson, 1997; Zeithaml, 1988; Dodds *et al.*, 1991; Yoo and Donthu, 1997; Washburn and Plank, 2002]. Aaker [1991] argued that a less famous brand tended to be less considered by consumers in terms of purchase. Erdem *et al.* [2004] discovered that the reliability of a brand would influence the brand choices of consumers since they tended to have prejudice that a brand represented the quality of the products. It is thus known that brand awareness and perceived quality and purchase intention are positively correlated.

Furthermore, previous literature also proved that brand awareness had critical influence on purchase intention [Dodds *et al.*, 1991; Grewal *et al.*, 1998; Washburn and Plank, 2002]. In the research on the influence of brand equity on the preference and purchase intention of consumers, Cobb *et al.* [1995] discovered that higher brand awareness tended to generate higher brand preference and purchase intention, which indicated that brand awareness was a crucial factor in the influence on the purchase intention of consumers.

In terms of purchase decision, the purchase intention of consumers for a more familiar brand is usually higher than that for a less familiar brand [Hoyer and Brown, 1990]. In addition, when the brand awareness of a brand becomes higher, and the price is slightly reduced, the profits that consumers perceived will be greater, and their purchase intention will significantly increase [Monroe and Krishnan, 1985; Kamins and Marks, 1991]. The aforementioned explains why the products with high brand awareness tend to have better quality evaluation and higher market share. Thus, we propose the following hypotheses:

H2a: Brand awareness has positive impact on perceived quality toward new products.

H2b: Brand awareness has positive impact on purchase intention toward new products.

3.3. *The influence of perceived quality on perceived value*

In light of the concept of perceived value, Monroe and Krishnan [1985] brought up the relational model of price, perceived quality, perceived sacrifice, perceived value, and purchase intention, which indicated that perceived value resulted from the balance between perceived quality and the perceived sacrifice of currency that consumers had to pay; when consumers had higher perceived quality on a product, there would be higher perceived value. Grewal *et al.* [1998] addressed that the acquisition value of consumers could be divided into the value obtained from products, usually referring to quality, and the transaction value of consumers. In which, the higher the product profits perceived by consumers is, the higher the perceived value would be. Namely, the higher the perceived quality of consumers is, the higher the perceived value would be. Furthermore, Dodds *et al.* [1991] and Chen and Dubinsky [2003] both discovered that perceived quality had significant positive influence on perceived value. Thus, we propose the following hypothesis:

H3a: Consumer-perceived quality has positive impact on perceived value toward new products.

3.4. *The influence of perceived value on purchase intention*

The establishment of the perceived value of consumers for a product is critical for enterprises to obtain competitive advantages [Parasuraman, 1997] and, meanwhile, it is an important index for the measurement of the repurchase intention of consumers [Parasuraman and Grewal, 2000; Woodruff, 1997; Petrick, 2002]. The perceived value of consumers for products or services will further generate purchase intention; purchase intention frequently depends on the difference between perceived profits and cost [Zeithaml, 1998; Dodds *et al.*, 1991; Grewal *et al.*, 1998]. A great deal of research found that the correlation between perceived value and purchase intention was positive, and purchase intention could effectively predict purchase behavior [Dodds *et al.*, 1991; Grewal *et al.*, 1998]. Monroe and Krishnan [1985] argued that the higher the perceived quality of consumers for a product is, the higher the perceived value for the product would be, and the purchase intention would be increased as well. In summary, perceived value and purchase intention are positively correlated. Thus, we propose the following hypothesis:

H3b: Consumer-perceived value has positive impact on purchase intention toward new products.

3.5. *The influence of perceived quality on purchase intention*

From both the model of perceived quality, perceived value, and purchase intention by Monroe and Krishnan [1985] and the causal relational model of value, perceived quality, and perceived value by Zeithaml [1988], it can be found that perceived quality influences the purchase intention of consumers; when the quality of a product perceived by consumers becomes higher, the perceived value increases, and the purchase intention also advances. Therefore, perceived quality has indirect influence on purchase intention. In addition, research, such as Dodds *et al.* [1991] and Olson and Jacoby [1977], indicated that perceived quality also had direct positive influence on purchase intention. Moreover, Rodoula [2006] discovered that perceived quality had simultaneous direct and indirect influences on purchase intention. In terms of practical experience, the higher the perceived quality of consumers is, the stronger the purchase intention will be. Thus, we propose the following hypothesis:

H4: Consumer perceived quality has positive impact on purchase intention toward new products.

4. Research Design

4.1. *Research framework*

To sum up the literature review and research hypothesis, this study establishes the research framework depicted in Figure 1, and a questionnaire survey and empirical study are used to verify the research hypothesis. This study uses two independent variables, the perceived innovation and brand awareness to explore whether they affect intermediary variables, perceived quality, and perceived value, and then dependent variable, purchase

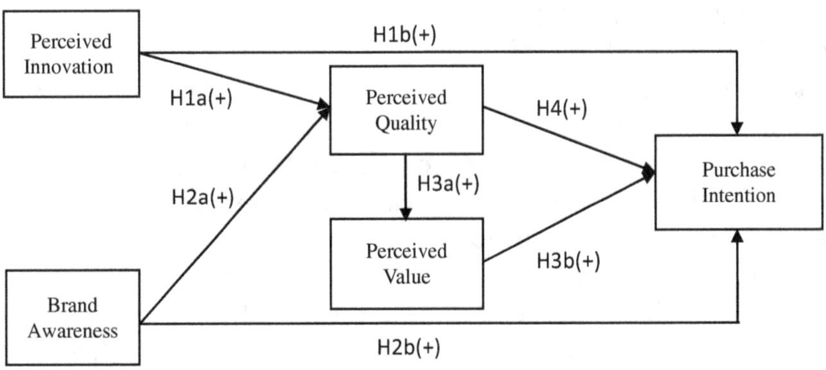

Figure 1. Research framework.

intention, following to discuss the interactive relationship among variables and their influence strengths.

4.2. *Questionnaire design*

The initial draft of the measurement variables was designed according to the theory and literatures review. The questionnaire was then generated in six parts; the first research perspective is perceived innovation, it including four subdimensions that are marketing innovation, product innovation, technological innovation, and service innovation [Rogers, 1962; Kwaku, 1995]. The second perspective is brand awareness, and there are four items [Berry, 2000; Aaker, 1996]. The third perspective is perceived quality, including five items [Petrick, 2002]. The fourth perspective, perceived value includes five items [Petrick, 2002]. The fifth dimension, purchase intention includes four questions [Dodds *et al.*, 1991]. The above five perspectives were measured by Likert 7-point scale, with a score of 1 meaning the lowest degree of agreement, and 7 meaning the highest degree of agreement. Final, the sixth part, the demographic variable including eight items, was measured by nominal scale (see Table 1).

4.3. *Pre-test and pilot*

To acquire an effective measurement scale, two-stage tests including pre-test and pilot were used to correct the questionnaire before the formal survey. In the pre-test, convenience sampling was used to select 30 cellphone users and conduct an in-depth interview. The results of these interviews revealed that some of the questions were not clear, and these sentences were revised for better description.

After correcting the questionnaire, convenience sampling was used again to select 50 people for pilot test, and these collected data was then used to analyze scale's reliability and validity. The pilot test results indicated that Cronbach's α of

Table 1. Measurement variables.

Measurement perspective		Items number	Definition	References
Perceived innovation	Marketing innovation	3	Consumers perceived innovation in the product or service and performance level.	Rogers [1962] Atuahene-Gima [1995] Weerawardena [2003] Johannessen *et al.* [2001]
	Product innovation	3		
	Technology innovation	3		
	Services innovation	3		
Brand awareness		4	Consumer cumulative memory strength toward the brand.	Berry [2000] Aaker [1996]
Perceived quality		5	Consumer perception on the quality of new product.	Petrick [2002]
Perceived value		5	Consumer's feeling about rationality and valuable toward a new product.	Petrick [2002]
Purchase intention		4	Consumer potential purchase intention toward a new product.	Dodds *et al.* [1991]
Demographic		8	Consumer's background.	N/A

measurement perspectives were all greater than 0.748; the factor loadings of all questionnaire items were greater than 0.5, and those meet the standards of reliability and validity [Nunnally, 1978]. It means this questionnaire scale has well reliability and convergent validity. Hence, the questionnaire was used for the formal survey of a large sample.

5. Research Results and Discussion

This study selected Taiwan's cellphone user as the sampling target. We used convenience sampling method to collect consumers' opinions about perceived innovation, brand awareness, perceived quality, perceived value, and purchase intention toward Apple's i-phone. We gathered data of 630 respondents during two months from June 2009 to July 2009. After excluding 35 invalid questionnaires, the valid questionnaires totaled 595 copies, for an effective return rate of 94.44%.

The respondents' demographics information is shown in Table 2. It shows that the respondents were 57.6% females; 57.6% of them were aged 21–30 years old and 21.7% were aged 31–40 years old; 53.9% of them got college degrees and 23.5% got master degrees; 24.5% of them were students, 21.2% were worked in service industry; most of them (30.3%) earned an average personal monthly income of US$301–900, followed by those who (28.1%) earned US$901–1,500; the job post of 54.5% of them was general staff.

This study uses structural equation modeling (SEM) to test the interrelated dependence relationships among multiple variables. Meanwhile, the fitness of model is judged by the rules of an excellent model's ratio of chi-square value to degree of freedom (χ^2/df), which should not be larger than 3 [Jöreskog and Sörbom, 1989]; the RMR and RMSEA should be smaller than 0.05, and GFI, AGFI, NFI, RFI, and CFI should be larger than 0.9 [MacIver and Carmines, 1981]. Table 3 shows that all fitness indexes reach or are close to the target value, therefore, the relationship model is acceptable.

Table 2. Demographic data of the sample.

Category items		Fre.	%	Items	Fre.	%
Gender	Male	252	42.4	Female	343	57.6
Residence	Northern of Taiwan	295	49.6	Eastern of Taiwan	6	1
	Central of Taiwan	228	38.3	The others	1	0.2
	Southern of Taiwan	65	10.9			
Age	Under 20	29	4.9	41–50	71	11.9
	21–30	343	57.6	More than 51	23	3.9
	31–40	129	21.7			
Education	Junior high school	11	1.8	Graduate school	140	23.5
	Senior high school	76	12.8	Doctor	47	7.9
	College	321	53.9			
Background or occupation	Worked in financial	21	3.5	Worked in commercial	38	6.4
	Worked in government	97	16.3	Worked in communication	19	3.2
	Worked in manufactory	78	13.1	Student	146	24.5
	Freelancer	17	2.9	The others	53	8.9
	Worked in service	126	21.2			

(Continued)

Table 2. (*Continued*)

Category items		Fre.	%	Items	Fre.	%
Personal monthly income (US$)	Under 300	150	25.2	1,501–2,100	48	8.1
	301–900	180	30.3	2,101–2,700	30	5.0
	901–1,500	167	28.1	More than 2,701	20	3.4
Job post	General staff	324	54.5	Entrepreneur	7	1.2
	Supervisor	38	6.4	Student	139	23.4
	Executive	19	3.2	The others	68	11.4

Table 3. Fitness indexes of the relationship model.

Fitness indexes										
χ^2	*df*	χ^2/df	*P* value	RMR	GFI	AGFI	NFI	RFI	CFI	RMSEA
641.123	323	1.985	0.000	0.097	0.931	0.907	0.945	0.931	0.972	0.041

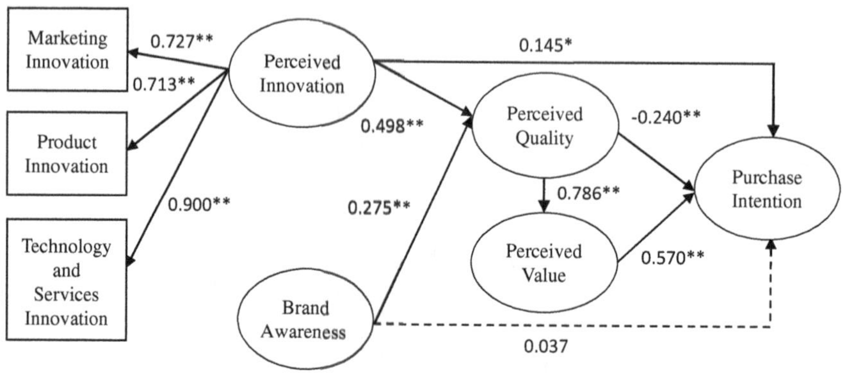

Figure 2. The relationship model.
Note: * p <0.05, ** p <0.01.

The SEM analysis (see Figure 2) show that the consumer's perceived innovation has a significant and positive impact on the perceived quality ($\gamma_{1a} = 0.498$, $P = 0.000$), this result supports H1a; the perceived innovation has a significant and positive impact on the purchase intention ($\gamma_{1b} = 0.145$, $P = 0.012$), this result supports H1b. Brand awareness has significant and positive impact on the perceived quality ($\gamma_{2a} = 0.275$, $P = 0.000$) that supports H2a; however, the brand awareness has no significant impact on the purchase intention ($\gamma_{2b} = 0.037$, $P = 0.439$), therefore H2b cannot be supported. Consumer perceived quality has a significant and positive impact on perceived value ($\beta_{3a} = 0.786$, $P = 0.000$) that supports H3a; consumer-perceived value has a significant and positive impact on purchase intention ($\beta_{3b} = 0.570$, $P = 0.000$) that supports H3b. Consumer-perceived quality has a significant and negative impact on purchase intention ($\beta_4 = -0.240$, $P = 0.000$), and these results cannot support H4.

5.1. *Influence effect*

Further, this study analyzes the influence effects of two independent variables, perceived innovation and brand awareness, on the dependent variable, purchase intention, that including the direct, indirect, and total effect (see Table 4).

Table 4. Analysis of the influence effects.

Independent variables	Mediator variables	Dependent variable	Direct effect	Indirect effect	Total effect
Perceived innovation		Purchase intention	0.145		
Perceived innovation	Perceived quality	Purchase intention		$0.498 \times (-0.240) = -0.119$	
Perceived innovation	Perceived quality Perceived value	Purchase intention		$0.498 \times 0.786 \times 0.570 = 0.223$	**0.249**
Brand awareness		Purchase intention	0.037		
Brand awareness	Perceived quality	Purchase intention		$0.275 \times (-0.240) = -0.066$	
Brand awareness	Perceived quality Perceived value	Purchase intention		$0.275 \times 0.786 \times 0.570 = 0.123$	**0.094**

The results show that the direct effect of perceived innovation is higher than brand awareness; after adding indirect effect, the total effect of perceived innovation on purchase intention (0.249) is still higher than brand awareness on purchase intention (0.094). The results indicate that consumer's perceived innovation has a more powerful drive than brand awareness on purchase intention of innovative product. Therefore, the main drive of consumer's purchase intention on innovative product is the consumer's perceived degree toward product's innovation. Thus, the purchase intention of innovative product will be reinforced if consumers perceive its innovation, that is an important finding of this study.

5.2. *Findings and discussion*

According to the research results (see Table 5), we found that the hypothesis H1a, consumer-perceived innovation, has a positive impact on perceived quality toward new products, and hypothesis H1b, consumer-perceived innovation, has a positive impact on purchase intention toward new products; both are supported and meet with the research findings of Kwaku [1995]. They found when customers understand product innovation, they will agree that some improvement has been done on a new product, then enhance their usage intention. Our research results also meet with the findings of Ottenbacher and Gnoth [2005], who pointed out that innovation can improve the quality of products or services; it also possessed indirect benefits in advancing corporate image, increasing the loyalty of customers, and attracting the purchase of potential customers. This study indicates that consumer-perceived innovation will raise both consumer-perceived quality and purchase intention, especially perceived quality.

The test result of Hypothesis H2a, brand awareness has a positive impact on perceived quality toward new products, is supported and meets with the findings of Wall *et al.* [1991] and Grewal *et al.* [1998]. They all proposed that the higher the

Table 5. The test results of hypotheses.

Hypothesis	Result	Coefficient
H1a: Consumer-perceived innovation has a positive impact on perceived quality toward new products.	Supported	0.498^{***}
H1b: Consumer-perceived innovation has a positive impact on purchase intention toward new products.	Supported	0.145^{**}
H2a: Brand awareness has a positive impact on perceived quality toward new products.	Supported	0275^{***}
H2b: Brand awareness has a positive impact on purchase intention toward new products.	Not supported	0.037
H3a: Consumer-perceived quality has a positive impact on perceived value toward new products.	Supported	0.786^{***}
H3b: Consumer-perceived value has a positive impact on purchase intention toward new products.	Supported	0.570^{***}
H4: Consumer-perceived quality has a positive impact on purchase intention toward new products.	Not supported	-0.240^{***}

Notes: **: $p<0.05$, ***: $p<0.01$

brand awareness, higher the consumer perceived quality. However, Hypothesis H2b, brand awareness, has a positive impact on purchase intention toward new products and has not been supported. Though Hoyer and Brown [1990] found that the purchase intention of consumers on a more familiar brand is usually higher than a less familiar brand; and Monroe and Krishnan [1985] and Kamins and Marks [1991] proposed when the brand awareness becomes higher and the price is slightly reduced, the consumers perceived profits will be greater, and their purchase intention will significantly increase. However, this study finds that even though well-known brands can give consumers a high perceived quality, they do not decide to buy just because of a higher brand awareness, especially, like a high-tech innovative product. Because Apple i-Phone was taken as the research object in this study, the general consumers in Taiwan recognize it is a well-known brand, but the price of i-Phone is very high at that time, so that they cannot make a purchase decision only based on high brand awareness. Thus, the perceived risk exists between brand awareness and purchase intention. Therefore, the impact of brand awareness on the consumer purchase intention cannot exist in the case of high-price innovative products.

Hypothesis H3a, consumer-perceived quality has a positive impact on perceived value toward new products, is supported and meet with the findings of Dodds *et al.* [1991] and Rao and Monroe [1989]. The hypothesis H3b, consumer-perceived value has a positive impact on purchase intention toward new products, also is supported and meets with Kotler [1994] and Petrick [2002]. It evidences that for the innovative and high-tech products with higher perceived quality, higher the perceived value will be, and stronger the purchase intention will be.

Hypothesis H4, consumer-perceived quality, has a positive impact on purchase intention toward new products, was not supported. Reasons maybe inferred from the survey which was performed from June 2009 to July 2009, at the time when the Apple i-Phone, with an expensive price, was just introduced into Taiwan; therefore, consumer cannot make purchase decision only according to the perceived quality of products. On the contrary, consumer may reduce their purchase intention due to perceived risks of innovative product, such as they have to change use method or the cost and sacrifice are higher than the product value [Ali *et al.*, 1995], and these reasons need to be further explored. Meanwhile, our study shows that consumer-perceived quality will directly influence the perceived value, then indirectly influence the purchase intention, that evidences the perceived value is a key mediator variable between perceived quality and purchase intention, since perceived value represents the gap evaluation between gain and expenditure. Therefore, the products with high quality and reasonable price will increase consumers' perceived value and cause their purchase intention.

6. Conclusions and Suggestions

6.1. *Conclusions*

This study is based on consumer perspective, to explore the relationship between perceived innovation, brand awareness, perceived quality, perceived value, and purchase intention. The results are acquisition through the literature review and empirical study. A structural equation model was constructed and the main findings are: (1) Consumer-perceived innovation has a significant and positive impact on the perceived quality and purchase intention, (2) Brand awareness has a significant and positive impact on the consumer-perceived quality, (3) Consumer-perceived quality has a significant and positive impact on perceived value, but negative impact on purchase intention, and (4) Consumer-perceived value has a significant and positive impact on purchase intention. These findings may contribute to the academic research and practical implications. There are five main findings in this research as follows:

(1) If consumers had strong perceived innovation for a new product, they would further perceive the changes and improvement of the quality of the product, but they would not thus generate purchase intention. Instead, their purchase intention would be strengthened by the advance of perceived value. Hence, the recognition of consumers for the value of innovative products is the key to increase purchase intention. In addition, the perceived value of new products depends on the difference between the payment and obtainment of consumers, so it is the goal of marketing experts to make consumers feel extra-value, or stimulate their feeling that something is worth buying. Furthermore, perceived innovation directly increases the purchase intention of consumers, which indicates it is a necessary marketing strategy to make consumers perceive the innovation of a product.
(2) The brand awareness of innovative product had significant positive influence on the perceived quality of consumers, indicating that the brand awareness brought consumers certain quality assurance, but it did not have significant influence on their purchase intention. According to this result, although a brand is internationally famous, it still cannot become the motive for purchasing a high-tech new product, so they will not decide to purchase the innovative product because it is brand-famous.
(3) Perceived quality has significant positive influence on perceived value, which indicates that the higher the perceived quality of consumers for a product is, the better the value of the product will be. Moreover, perceived value also has significant positive influence on purchase intention, namely, the higher the perceived value of consumers is, the stronger the purchase intention will be. This is the major method for stimulating consumers to purchase.

(4) Perceived quality and purchase intention were significantly and negatively correlated, which was not consistent with the hypotheses in this study. It is possible that consumers only thought the quality of the new product of a famous brand should be good, but since they did not use the product personally, and the price was higher, so the extra price for the innovative product could not be offset through quality benefit.
(5) According to the effectiveness analysis, the variable influencing the total effect of the purchase intention of consumers the most was perceived innovation, whose value was 0.249. The other influential variable was brand awareness, but the influence value on the total effect of the purchase intention was only 0.094. This indicated that if manufacturers would like to strengthen the purchase intention of consumers for their innovative products, it would be better to enhance the perception of consumers for the innovative methods of new products in marketing, products, technologies, or services.

6.2. *Managerial implications*

According to this study, the major factor in the advance of purchase intention lies in whether or not consumers can perceive the innovation of a product, which indicates that the purchase intention for a new product will be increased if consumers can perceive the innovation. Therefore, it is a necessary marketing strategy for manufacturers to spread the information about innovation.

Since Apple i-Phone 3G was introduced in Taiwan in September, 2008, it tended to belong to highly innovative products for the consumers investigated at that time. It was thus necessary for these consumers to have more information and consider other factors to make the purchase decision. Consequently, the manufacturers of highly innovative products should not only promptly convey the innovative information on products to consumers through multiple channels, but also hold the trial activities of new products.

Since the price of i-Phone belonged to the high-price products just introduced to the market in the investigation period of this study, the purchase intention of consumers might thus reduce. Hence, it is suggested that in the initial stage when an innovative product is introduced, the manufacturer should focus on how to make consumers feel the innovation is worth purchase or the new product is lofty and unique, so that the purchase intention of consumers will be stimulated still more, and the sales of the product will be thus increased. When the innovative product enters the growing stage, the manufacturer should adjust the price and combine it with promotions to attract the potential consumers who recognize its quality, so that the perceived sacrifice of consumers will be reduced, and the perceived value will be advanced to further increase the purchase intention.

6.3. *Limitations and future research*

Apple i-Phone was taken as the research object in this study. Consequently, proper evaluation is possibly necessary if the research result is going to be applied to other products. For example, it is necessary to consider whether the product is a durable or a non-durable, or whether or not there may be different reactions if the target customer samples are different.

Furthermore, there are a lot of external clues for consumers to evaluate perceived quality and purchase intention. However, only brand awareness and perceived innovation were investigated in this study, it did not include all the influential factors. More influential variables such as the source country, perceived risk, and price of a product, can be included in the future research to further complete the research.

Different target groups or industries may also differentiate research results, so it is suggested that following research may investigate other industries to test and verify again the practicability of the conceptual framework of this study.

The respondents are from Taiwan's cellphone user group; however, Taiwan is a developed country with a wonderful infrastructure in telecommunication. The behavior of the cellphone users in Taiwan might be different from other countries. Therefore, the research results may not be able to represent the general applicability. The future research can be expanded to different countries using the proposed model, in order to evidence the effectiveness of the relationship model.

With the introduction of so many sophisticated cellphones in the past years, using only i-Phone might not be appropriate. Inclusion of several leading brands with wonderful brand awareness could enhance the value of the paper such as: Samsung Galaxy, Motorola Droid, etc. The future research can add other brands in order to evidence the general effectiveness of the hypotheses and proposed model.

References

Aaker, D. A. (1991). The value of brand equity. *Journal of Business Strategy*, **13**(4):27–32.

Aaker, D. A. (1996). *Building Strong Brands*. The Free Press, NY.

Ali, A., Krapfel, R. and LaBahn, D. (1995). Product innovativeness and entry strategy: Impact on cycle time and break-even time. *Journal of Product Innovation Management*, **12**(1):54–69.

Atuahene-Gima, K. (1995). An exploratory analysis of the impact of market orientation on new product performance: A contingency approach. *Journal of Product Innovation Management*, **12**(4):275–293.

Berry, L. L. (2000). Cultivating service brand equity. *Journal of the Academy of Marketing Science*, **28**(1):128–137.

Betz, F. (1993). *Strategic Technology Management*. McGraw-Hill, NY.

Bhuian, S. N. (1997). Marketing cues and perceived quality: Perception of Saudi consumers toward products of the U.S., Japan, Germany, Italy, U.K. and France. *Journal of Quality Management*, **1**(2):217–235.

Biswas, A. (1992). The moderating role of brand familiarity in reference price perceptions. *Journal of Business Research*, **25**(3):251–262.

Bitta, A. J. D., Monroe, K. B. and Mcginnis, J. M. (1981). Consumer perceptions of comparative price advertisement. *Journal of Marketing Research*, **18**(3):416–427.

Bredahl, L., Brunso, K. and Grunert, K. G. (2004). Consumer perception of meat quality and implications for product development in the meat sector-A reviews. *Meat Science*, **66**:259–272.

Brentani, U. (2001). Innovative versus incremental new business services: Different keys for achieving success. *Journal of Product Innovation Management*, **18**(3):169–187.

Chen, Z. and Dubinsky, A. J. (2003). A conceptual model of perceived customer value in E-commerce: A preliminary investigation. *Psychology and Marketing*, **20**(4):323–347.

Cobb, W. C. J., Ruble, C. A. and Donthu, N. (1995). Brand equity, brand preference, and purchase intent. *Journal of Advertising*, **24**(3): 25–48.

Danneels, E. and Kleischmidt, E. J. (2001). Product innovativeness from the firm's perspective: Its dimensions and their relation with project selection and performance. *Journal of Product Innovation Management*, **18**(6):357–373.

Dodds, W. B., Monroe, K. B. and Grewal, D. (1991). Effect of price, brand and store information on buyers' product evaluations. *Journal of Marketing Research*, **28**(3):307–319.

Erdem, T., Ying, Z. and Ana, V. (2004). Performance of store brands: A cross-country analysis of consumer store-brand preference, perceptions, and risk. *Journal of Marketing Research*, **41**(1): 86–100.

Fishbein, M. and Ajzen, I. (1975). *Belief Attitude Intention and Behavior: An Introduction to Theory and Research*. Addison-Wesley, Reading, MA.

Garvin, D. A. (1983). Quality on the line. *Harvard Business Review*, **61**: 65–73.

Grewal, D., Monroe, K. B. and Krishnan, R. (1998). The effects of price-comparison advertising on buyers' perceptions of acquisition value, transaction value and behavioral intentions. *Journal of Marketing*, **62**(2):46–59.

Horn, D. and Salvendy, G. (2006). Consumer-based assessment of product creativity: A review and reappraisal. *Human Factors and Ergonomics in Manufacturing*, **16**(2):155–175.

Hoyer, W. D. and Brown, S. P. (1990). Effect of brand awareness on choice for a common repeat-purchase. *Journal of Consumer Research*, **17**(2):141–148.

Jacoby, J. and Olson, J. C. (1997). *Consumer Response to Price and Attitudinal, Information Processing Perspective*, Eds. Y. Wind and P. Greenberg, American Marketing Association, Chicago, pp. 73–86.

Johannessen, J. A., Olaisen, J. and Olsen, B. (2001). Mismanagement of tacit knowledge: The importance of tacit knowledge, the danger of information technology, and what to do about it. *International Journal of Information Management*, **21**:3–20.

Johne, A. (1999). Using market vision to steer innovation. *Tech- innovation*, **19**(4):203–207.

Jöreskog, K. G. and Sörbom, D. (1989). LISREL 7 Use's Reference Guide. Chicago: Scientific Software.

Kamins, M. A. and Marks, L. J. (1991). The perception of kosher as a third party certification claim in advertising for familiar and unfamiliar brands. *Journal of Academy of Marketing Science*, **19**(3): 177–185.

Keller, K. L. (1993). Conceptualizing, measuring and managing customer-based brand equity. *Journal of Marketing*, **57**(1):1–22.

Kotler, P. (1994). *Marketing Management: Analysis, Planning, Implementation, and Control*, 8th edn., Prentice-Hall International Inc.

Krugman, H. E. (1965). The impact of television advertising: Learning without involvement. *Public Option Quarterly*, **29**:349–356.

Kwaku, A. G. (1995). An exploratory analysis of the impact of market orientation on new product performance: A contingency approach. *Journal of Product Innovation Management*, **12**:275–293.

Lane, V. (2000). The impact of ad repetition and ad content on consumer perceptions of incongruent extensions. *Journal of Marketing*, **64**(1):80–91.

MacIver, J. P. and Carmines, E. G. (1981). Unidimensional Scaling. California: SAGE.

Monroe, K. B. (1973). Buyers' subjective perceptions of price. *Journal of Marketing Research*, **10**(1):70–80.

Monroe, K. B. (1991). The value concept and relationship marketing. *European Journal of Marketing*, **30**(2):19–30.

Monroe K. B. and Krishnan, R. (1985). *The Effect of Price on Subjective Product Evaluation in Perceived Quality: How Consumer View*

Stores and Merchandise, Eds. J. Jacoby and J. C. Olson. Lexington Books, Lexington, MA, pp. 209–232.

Moorman, C., Zaltman, G. and Deshpande, R. (1992). Relationships between providers and users of market research: The dynamics of trust within and between organizations. *Journal of Marketing Research*, **29**:314–328.

Morwitz, V. G. and Schmittlein, D. (1992). Using segmentation to improve sales forecasts based on purchase intent: Which "intenders" actually buy? *Journal of Marketing Research*, **29**:391–405.

Nunnally, J. C. (1978). *Psychometric Theory*, 2nd edn., McGraw-Hill, NY.

Ottenbacher, M. and Gnoth, J. (2005). How to develop successful hospitality innovation. *Cornell Hotel and Restaurant Administration Quarterly*, **46**(2):205–222.

Parasuraman, A. (1997). Reflections on gaining competitive advantage through customer value. *Journal of the Academy of Marketing Science*, **25**(2):154–161.

Parasuraman, A. and Grewal, D. (2000). The impact of technology on the quality-value-loyalty chain: A research agenda. *Journal of the Academy of Marketing Science*, **28**(1):168–174.

Petrick, J. F. (2002). Development of multi-dimensional scale for measuring the perceived value of a service. *Journal of Leisure Research*, **34**(2):119–134.

Prajogo, D. I. and Sohal, A. S. (2006). The integration of TQM and technology R&D management in determining quality and innovation performance, *Omega*, **34**(3):296–312.

Rao, A. R. and Monroe, K. B. (1989). The moderating effect of prior knowledge on cue utilizations. *Journal of Consumer Research*, **15**: 253–264.

Rodoula, T. (2006). The role of perceived product quality and overall satisfaction on purchase intentions. *International Journal of Consumer Studies*, **30**(2):207.

Rogers, E. M. (1962, 1995). *Diffusion of Innovations*, 1st, 3rd and 4th edn. The Free Press, NY.

Samiee, S. (1994). Customer evaluation of products in a global market. *Journal of International Business Studies*, **25**(3):579–604.

Schepers, J., Schnel, R. and Voom, P. (1999). From ideas to business-how Simon bridges the innovation gap. *Journal of Research Technology Management*, **3**(42):26–31.

Schumpeter, J. A. (1934). *The Theory of Economic Development*. Harvard University Press, Cambridge.

Shelton, R. (2009). Integrating product and service innovation. *Journal of Research Technology Management*, **52**(3):38–44.

Shimp, T. A. and Bearden, W. (1982). Warranty and other extrinsic cue effects on consumer risk perceptions. *Journal of Consumer Research*, **9**:38–46.

Steenkamp, J. B. (1990). Conceptual model of the quality perception process. *Journal of Business Research*, **21**:309–333.

Veryzer, R. W. J. (1998). Discontinuous innovation and the new product development process. *Journal of Product Innovation Management*, **15**(4):304–321.

Wall, M., Liefeld, J. and Heslop, L. A. (1991). Impact of country of origin cues on consumer judgments in multi-cue situations: A covariance analysis. *Journal of the Academy of Marketing Science*, **19**(2): 105–113.

Washburn, J. H. and Plank, R. E. (2002). Measuring brand equity: An evaluation of a consumer-based brand equity scale. *Journal of Marketing Theory and Practice*, **10**(1):46–63.

Weerawardena, J. (2003). The role of marketing capability in innovation-based competitive strategy. *Journal of Strategy Marketing*, **11**(1):15–35.

Woodruff, R. B. (1997). Customer value: The next source for competitive edge. *Journal of the Academy of Marketing Science*, **25**(2): 139–153.

Yoo, B. and Donthu, N. (1997). Developing and validating a consumer-based overall brand equity scale for Americans and Korean: An extension of Aaker and Keller's conceptualizations. *Paper presented at 1997 AMA Summer Educators Conference*, Chicago.

Zeithaml, V. A. (1988). Consumer perceptions of price, quality, and value: A means-end model and synthesis of evidence. *Journal of Marketing*, **52**(3):2–22.

Zeithaml, V. A. and Bitner, M. J. (2000). *Service Marketing: Integrating Customer Focus across the Firm*, 2nd edn. McGraw-Hill Companies.

Chapter 4

International Perspectives: Case of India

Dharmesh Gupta, Ruchita Gupta, Karuna Jain, and Kirankumar S. Momaya

1. Introduction

Telecommunication (telecom) is one of the earliest forms of technology-based services, which has recently become the center of innovation. Mobile communications, in particular, is one of the most striking technology-based service innovation in the last 30 years. Worldwide mobile subscribers have now reached 6.91 billion in December 2014, with most subscribers in developing countries (5.4 billion) [ITU, 2014]. Mobile phone usage has surpassed desktop internet usage [Lipsman, 2014]. The evolution and adoption of mobile telephony has been overwhelming in India. Competition between service providers, competition with fixed-line telephony, and government intervention have contributed to the fast diffusion of mobile telephony in India [Gupta and Jain, 2012], making mobile density reach 75.5% in 2014 from 4% in 1995 [TRAI, 2014]. In recent past, it was observed that the use of mobile telephony has direct impact on the economic growth and social development of a country [Abraham, 2007].

India is a growing market in Southern Asia with more than 12 operators and 937 million mobile subscribers [TRAI, 2014]. While there is a significant growth in mobile density, there exists a gap between urban and rural mobile density (urban — 146%; rural — 44%) [TRAI, 2014] as shown in Figure 1.

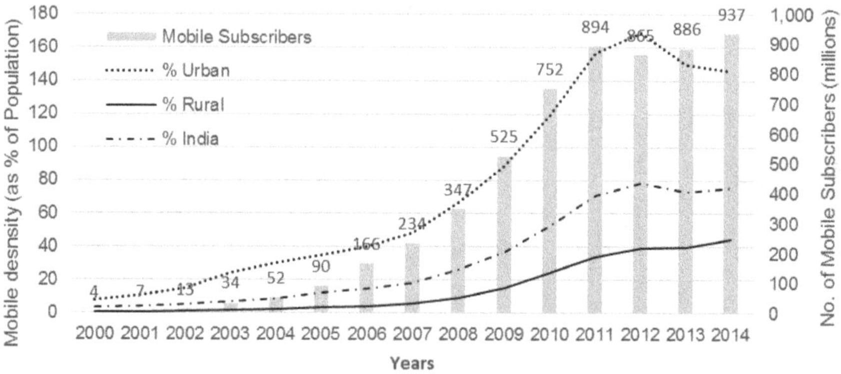

Figure 1. Growth of mobile telephony in India [TRAI, 2014].

Mobile phones are used for various services such as short message service (SMS), multimedia message service (MMS), playing music, information on demand (IOD), mobile banking, mobile internet, email, and playing games. Today, the telecom industry stands on the cusp of another revolution called mobile VAS, which is the new engine of growth for mobile operators. The telecom operators are facing a dilemma, on the one hand, they need to develop new businesses such as new mobile data applications and, on other, they do not have sufficient financial resources to invest in the entire infrastructure while keeping pace with emerging opportunities. They have adopted product-platform-based approach to accentuate the growth of mobile (VAS) services. Largely, this approach is found in practice for business-to-consumer (B2C) services [Eduardo and Sato, 2014].

Furthermore, internet users in India have grown from 7 million in 2001 to 302 million in 2014 [BCG and IAMAI, 2014] as shown in Figure 2. Urban India has 190 million internet users (63%), while rural India has 112 million (37%) users, with rural-to-urban user's ratio as 0.59 [IAMAI *et al.*, 2014].

The growth of the two technologies is leading to convergence at the device as well as service level. The technological convergence as a special case of emerging disruptive innovation in information and communication technologies (ICT)

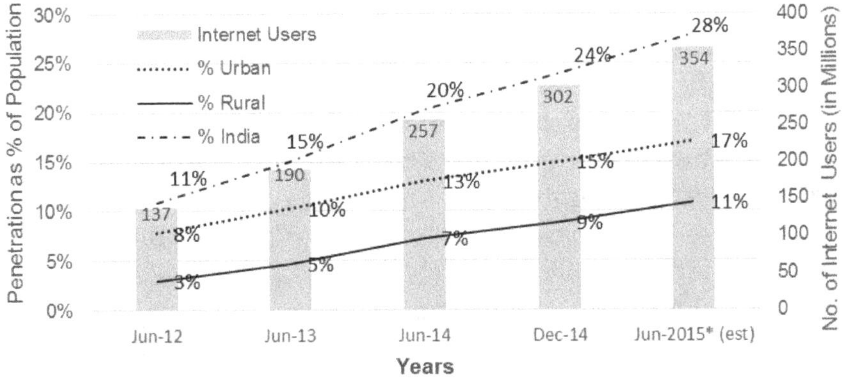

Figure 2. Growth of internet users in India [IAMAI *et al.*, 2014].
Note: *Expected number.

industry has been studied by Hacklin *et al.* [2005]. The interplay of mobile and internet technologies is blurring the boundaries between traditional telecom and Internet, creating new business models and market opportunities for mobile service innovation and service delivery. With this convergence, services like online banking, Internet protocol television (IPTV), data analytics as a service (big data platform), or smart devices (Internet of things (IOT)) in remote monitoring, health-care diagnostics, home-automation, manufacturing, logistic, and education are also converging at the mobile devices. This presents an opportunity to provide an efficient and cost-effective platform for connecting people with relevant services.

There are two approaches the telecom operator can use to offer mobile VAS — walled garden (where operator tries to offer everything) and open. To maintain their position, operators have sought to develop "walled garden" portals to ensure that content creators are kept well away from end customers [Olla and Patel, 2002; Peppard and Rylander, 2006]. This strategy made it tough for content providers to easily connect to telecom operator to make their content available, and those that do are also discontented with the revenue-sharing arrangements [Peppard and Rylander, 2006]. Until a couple of years ago, telecom operators in India favored focusing on their walled gardens. Instead of providing pervasive access to the Internet, they had their own wireless access protocol (WAP) portals through which — along with voice- and SMS-based services, they provided VAS. These emerging services have pushed telecom operators and equipment providers from wall-gardened realm into open market, leading to the deconstruction of telecom value chain [Li and Whalley, 2002; Pil and Holweg, 2006]. Also, technological innovation has been the main driver for the development of new mobile VAS. Mobile VAS has grown significantly in terms of variety, number, and revenue. Table 1 presents the gap in the penetration of the essential services like healthcare, education, and banking between India and the developed world. The large gap presents an opportunity for mobile operators to provide an efficient and cost-effective platform for connecting people and services. Innovations in mobile VAS offer a significant potential for inclusive growth, bridging digital divide, and inequalities, in developing countries like India.

The three key trends in telecom industry in last few years — (a) growth in number of mobile subscribers, (b) increase in number and variety of services, and (c) locus of telecom service development (as many of the emerging services are developed outside the realm of conventional telecom service providers (operator or technology vendors)), are motivated to under-

Table 1. Touchpoints per 1,000 people using all available modes of reaching out in India (2009).

Country	Health care (doctors)	Education (school)	Financial (Bank)
India	1.9	24.9	0.07
US	12.5	70.4	0.32
Japan	11.6	52.2	0.13
Germany	11.5	70.5	0.55

Source: Deloitte and ASSOCHAM [2011].

stand the process of service innovation in new value web ecosystem, crossing the traditional sectoral boundaries. Is the existing telecom service delivery framework adequate for such services? In the multiple-partner scenario, who owns the customer and take the ownership of end-to-end service ownership? How the role of telecom operator is changing?

Thus, the objective of this research is to understand how telecom operators in India are engaging in mobile VAS development, delivery, and operations. Also, the study aims to assess the applicability of the selected framework in the evolving context of mobile service delivery. In this research, mobile VAS in both B2C and business-to-business (B2B) segments, introduced by an Indian telecom operator are analyzed. These services have been analyzed from three perspectives — (a) process of service innovation, (b) assessing and mapping technological competencies and capabilities required for new service development, and (c) service operations (production). This study focuses on the interplay between technological innovation and service innovation, and analyzes cases from service management perspective using the enhanced Telecom Operations Map (eTOM).

In the following section, review of the literature is presented with the changing scenario of mobile services in India, innovation models, and operations framework for telecom service delivery. The research methodology is then explained in the next section. The chapter then presents the select case studies. The process of each mobile VAS service innovation is analyzed and mapped using eTOM. Findings from the case analysis are then presented. Emerging issues are then discussed, and conclusions are drawn.

2. Literature Review

Based on the objective of the research to study the role of telecom operators in India for mobile VAS development, delivery, and operations, literature is reviewed in different dimensions. Different models of product and service innovations and evolving concepts in the service management literature are studied. Further, operations frameworks for telecom service delivery are also reviewed.

2.1. *Innovation models*

Existing innovation models in the literature make distinctions in technology/product innovation [Rothwell, 1994; Tether and Howels, 2007] and service innovation [Barras, 1986; Song *et al.*, 2009; Hertog *et al.*, 2010]. This difference stems from the specific characteristics of services — intangibility, co-production with customers, simultaneity or inseparability, heterogeneity, and perishability [Papastathoulou and Hultink, 2012]. Further researchers have defined service as an application of knowledge and competencies for the benefit of another entity, which makes it the basis of any economic or social exchange [Vargo *et al.*, 2008]. Researchers critically analyzed the differentiating services characteristic (intangibility, heterogeneity, inseparability, and perishability) and argued that there is a dichotomy of characteristics of products and services. It is unfair to generalize these characteristics and differentiate the product and services by these alone [Lovelock and Gummesson, 2004; Vargo and Lusch, 2004]. Technology-based manufacturing businesses are adopting service-oriented thinking and are increasingly placing more focus on value co-creation to generate new value proposition and business models, thereby promoting servitization in technology-based companies [Ulaga and Reinartz, 2011; Shirahada *et al.*, 2015]. However, telecom services have always been delivered as "servitization". Moreover, the service-dominant (S-D) logic places the customer at center stage, such that the customer is always a co-producer in service exchange. The value co-creation process involves the supplier creating superior value propositions, with customers determining value when a good or service is consumed. It results in greater opportunities for co-creation and results in mutual benefits (or "value") being received by the supplier by way of revenues, profits [Payne and Frow, 2005].

These characteristics affect the development process making the service innovation process unique [de Brentani, 1991]. Therefore, to address the unique aspects of services, the need for new service innovation model is asserted [Gadrey *et al.*, 1995; Coombs and Miles, 2000; Djellal and Gallouj, 2001; Menor *et al.*, 2002]. Initially, the service innovation models based on resource-based view (RBV) focused on integrating resources and static capabilities for innovation. However, RBV does not address how firms integrate resources and capabilities in a dynamic environment [Lawson and Samson, 2001]. Later, dynamic capability framework (DCF) evolved, which focused on the reconfiguration of internal and external competencies in a rapidly changing environment [Teece *et al.*, 1997; Helfat and Peteraf, 2003]. Open service innovation models are being applied in the telecom services to access external competencies to enhance the speed of the innovation process [Nesse, 2009; Birudavolu and Nag, 2011, 2015; Bigliardi *et al.*, 2012].

Further, the characteristic-based approach [Lancaster, 1966] to innovation is an integrative approach to goods and services representation. In this approach, a product (good or service) is considered as a set of functional characteristics. These characteristics of the product are seen from the end-user perspective, e.g., in the case of a car, its size, performance, comfort, and safety features are seen from end-user perspective [Saviotti and Metcalfe, 1984; Gallouj and Weinstein, 1997]. Figure 3 illustrates the characteristics-based representation of a product, where vector [TC] represents the technical characteristic of the underlying product. Vector [C] indicates a set of competencies of suppliers or partners. Development and delivery of a new service could, therefore, be defined as a process of simultaneous interplay of technical characteristics and competencies (internal or external) to create new characteristics, [Y], representing the final service functional characteristics of new

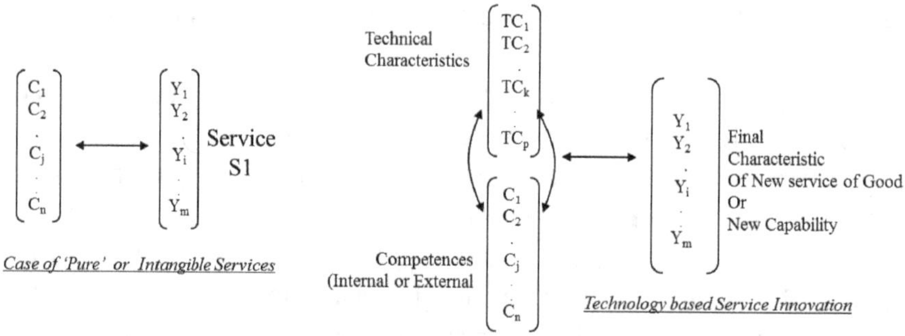

Figure 3. Representation of a product or service as a system of characteristics and competences [Gallouj and Weinstein, 1997].

service, or the creation of new TC, or acquiring new [C]. Different types of innovations (radical, incremental, modular, and architectural) and degree of their novelty are the results of the dynamic application of [TC] and [C]. Thus, the final characteristics that are obtained through a certain combination of technical characteristics and competency, which describe the organization's tangible and intangible systems, are used in the production of the service. For example:

(1) A physiotherapist treating a patient is an example of service, wherein medical competency is applied to offer a service.
(2) A database system to provide a wide variety of IOD is a service, wherein database is the technical characteristic and ICT integration is the competency applied to provide service on demand. Further, the extension of this example is designing of a service based on a new way of processing and utilizing information, or combining one or more of existing databases (such as Big Data Analytics). Here database and IOD is the technical characteristic and computation is the competency.
(3) IOT service such as remote operation of a device (enabling or disabling), wherein device and communication platform

are technical characteristics and system integration is the competency.

In all these examples, relevant competencies are employed with technical characteristics to create a new service or product. The degree of novelty depends on dynamics of such interactions and intensity of use.

2.2. *Operations framework for telecom service delivery*

In addition, as the interaction between new service development and service delivery is high [Nijssen *et al.*, 2006], the management of telecom service and network infrastructure encompasses a number of integrated systems in which operations support system (OSS) and business support systems (BSS) form a crucial part. For many years, telecom operators followed the telecommunications management network (TMN) model recommended by the International Telecommunications Union (ITU) to develop their network management strategies. The convergence scenario is changing the traditional OSS/BSS and a number of business process models/frameworks have evolved, such as — eTOM, Information Technology Infrastructure Library (ITIL), Control Objectives for Information and Related Technology (COBIT). Korzachenko and Getman [2010] did the comparison of these frameworks for telecom enterprise on six parameters — Openness, Systematic, Business or Technology orientation, Elaboration depth, Appreciation of information technology (IT), and Specificity and Appreciation of telecom industry specificity. eTOM was suggested as the effective business process framework that allows telecom companies to provide modern, high quality, and cost-competitive services [Korzachenko and Getman, 2010], and has gained popularity in various countries like Taiwan and Ukraine [Wong, 2009; Chang, 2011]. While the TMN focuses on network management, eTOM focuses on service and customer management.

The eTOM is targeted for both intra-enterprise use and inter-provider process integration. It uses hierarchical decomposition which describes all operator activities in a standard way. At the overall conceptual level, the eTOM can be viewed as having three major process areas: (1) Strategy, infrastructure, and product (SIP) (covering planning and lifecycle management associated with development and delivery, (2) Operations (covering the core of operational management — fulfillment, assurance, and billing (FAB)), and (3) Enterprise management (covering corporate or business support management), and is shown in Figure 4.

Each process area is further divided into vertical and horizontal subprocess groups overlapping each other (called level 2 grouping). The level 2 decomposition provides the end-to-end processes view, which play a critical role in support for customers

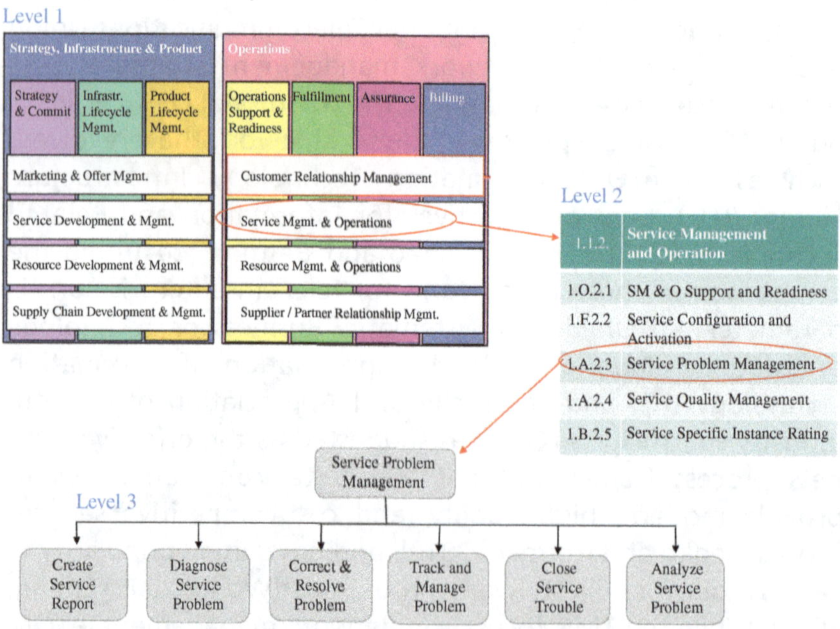

Figure 4. eTOM business process framework [TM forum and ITU-T M.3050, 2007].

to manage the business and quality of service (QoS) end-to-end. For example, vertical level 2 grouping for operation processes are operation and support readiness, FAB, and revenue management. Horizontal grouping for operation processes consists of customer relationship management (CRM), service management and operation, resource management and operations, and supplier/partner relationship management. Figure 4 shows seven vertical process groupings and eight horizontal subprocesses under SIP and operation processes. At the intersection of horizontal and vertical processes, subprocesses are further defined, called level 3 subprocesses.

2.3. *Analysis of literature review*

It is evident from the analysis of literature that the underlying fundamental dimensions for product and services are the same; however, intensity and degrees of these may differ. Gadrey and Gallouj [2002] mentioned that innovation is predominantly a service activity. Technology-based service innovation is essentially an interplay between technology and service focusing on creating value for the end user. Service innovation is considered critical for the survival of the firm and economic growth of the nation. However, research framework for the management of service innovation and new service development remains scarce [Feri, 2008; Moller *et al.*, 2008; Kim *et al.*, 2015].

With increasing speed of technological change and competitive dynamics, it is difficult for firms to maintain capabilities in all potential technical and market areas [Lawson and Samson, 2001]. It requires firms to stretch beyond the traditional boundaries widening the opportunity and access to key resources. Also, higher-order integration capability is critical to stimulate innovation [Lawson and Samson, 2001]. Therefore, a more holistic approach is required towards service innovation and development. Despite literature on innovation models and business process frameworks, there is a lack of empirical evidence to verify existing and popular eTOM in the practical and changing scenarios, especially in the context of India. Here, the authors take the opportunity to explore this gap in the Indian context.

3. Research Objectives and Methodology

The key purpose of the research is to understand how telecom operators in India are engaging in mobile VAS development, delivery, and operations. Further, the research aims to consider the applicability of the selected framework in changing context of mobile service operations and delivery. A case-study research design is adopted to understand the development of mobile service innovations characterized by complexity and contextual features [Yin, 2013]. Case studies help in the enhanced understanding of a phenomenon in its natural context by providing insights on the how-and-why aspects [Yin (2013)], especially when constituent dimensions are difficult to measure quantitatively [Caerteling *et al.*, 2008]. Sample cases from specified population should be selected on the basis of judgment [Glaser and Strauss, 1967]. Criteria like possible replication, theory extension, and contrary findings are considered [Yin, 1994]. Consequently, judgmental sampling is adopted for selecting the three service innovation cases. Cases are selected carefully to represent the spectrum of services evolution from basic service like information-on-demand (created within the telecom service providers' realm) to advanced services like WhatsApp (created outside the telecom service providers' realm). The select telecom operator "*X*" has been offering an entire full range of services covering the diversity of technology and spread across the country.

Three mobile VAS (ring back tone (RBT), smart energy meter (SEM), and WhatsApp messenger) introduced by the Indian mobile operator are selected for the research as shown in Figure 5. The cases are selected based on four distinctive criteria — (i) Service ownership, (ii) Customer type — retail/direct customer

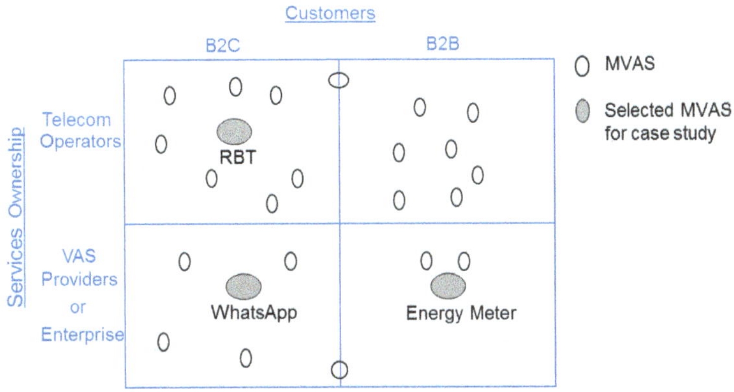

Figure 5. Mobile VAS selected for multicase study.

(B2C) or enterprise (B2B), (iii) Popularity, adoption and revenue contribution of the service, and (iv) Authors' involvement in the service development and implementation. A qualitative approach is adopted with an effort to generate in-depth and rich descriptions [Denzin and Lincoln, 1994] about the development and management of mobile service innovations which can be of considerable utility for practitioners. Open-ended interviews were conducted with the team of experts involved in the development and management of the select service innovations. A co-author of the paper was also involved in the development of services introduced by the Indian telecom operator. Eight interviews were conducted with key officials involved in value addition activities associated with the three case studies. The duration of interviews ranged from 60 to 180 minutes. Respondents were provided the discussion guide before the scheduled interview so that they could prepare themselves. Data were analyzed in three steps: data reduction, data display, and drawing conclusions [Berkowitz, 1997; Miles and Huberman, 1994].

For technology-based service development and service innovation, the authors adopts Gallouj's characteristics model and consider it as the basis of analysis for technology-based service development in telecom domain. The study illustrates the development of mobile service innovations on integrated product–service framework [Saviotti and Metcalfe, 1984; Gallouj and Weinstein, 1997; Lancaster, 1996] and its management by mapping processes on eTOM. Business processes for each of the mobile VAS service case are mapped to eTOM business process framework (level 2/3) for SIP and FAB.

4. Select Case Studies and their Analysis

The service innovation process includes four stages — ideation, development, implementation, operations, and delivery. Development of the new service necessitates new capability development (through different existing capabilities and competency within the firm or from multiple partners) as shown in Figure 6. The cases have been analyzed from three perspectives — (a) Process of service innovation, (b) Assessing and mapping competencies and technological capabilities required for new service development (Block A in Figure 6), and (c) Service operations and delivery. Service operations and delivery have been analyzed by mapping the operational processes with the existing process frameworks. The analysis of the three

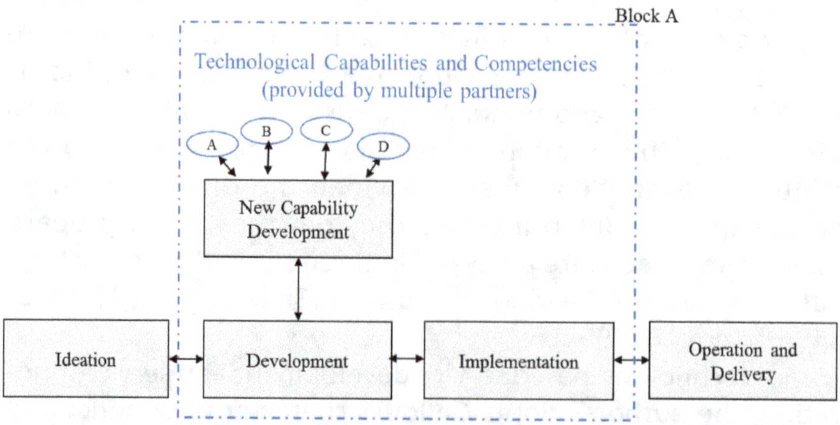

Figure 6. Process of service innovation.

cases on mobile VAS is now presented with the perspective of their development as well as the operational process of the service innovation. The select cases represent the majority of the spectrum of mobile VAS being offered by telecom operators.

4.1. *Process of service innovation and mapping technological capabilities*

Case 1: Ring Back Tone

Ideation Stage

RBT is a musical ring tone of choice (preselected and stored in the service provider's network system), which a caller hears while waiting for a phone to be answered. RBT — a B2C service was launched in the year 2007 by the Indian telecom operator "*X*" with the motive to be recognized as an innovative brand, capture the market, and enhance revenue, thereby illustrating the technology-push approach of service innovation. RBT is one of the most successful mobile VAS launched in the Indian market. The flow of information in RBT service was ideated as shown in Figure 7.

i. Mobile subscribers subscribe to RBT service which is be then tagged for RBT call forwarding in Home Location Register (HLR). With this, all calls are thus forwarded/ directed toward RBT platform.
ii. RBT subscribers call up an interactive voice response (IVR) system, select music/tunes of their choices from RBT content servers. These identified musical tunes are flagged in RBT platform, which will be played when someone calls up on mobile number of the RBT subscribers.
iii. When any other mobile user or any public switched telephone network (PSTN) customer makes a call to mobile number of an RBT subscriber, the modified functionality in mobile switching center (MSC) (for call forwarding, call connect, and release functionality for RBT calls) identifies

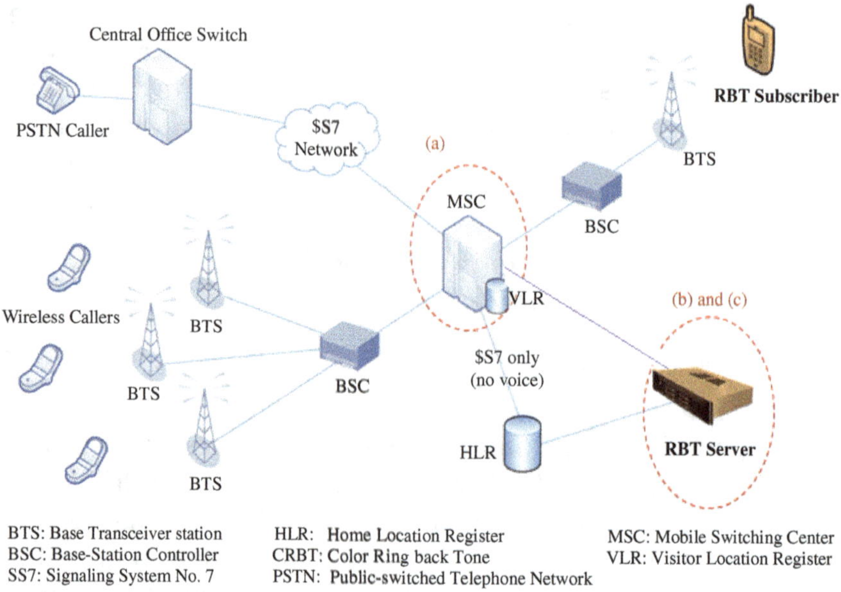

Figure 7. Information flow in RBT service.

that call for an RBT subscriber and forwards the call to RBT platform.

iv. RBT application receives the call on Signaling System No. 7 (SS7) interface, connects the call to the content server, and plays the preselected musical tune to the caller.

v. While playing music to the caller, RBT application makes another call to the called RBT subscriber using SS7 signaling. Again, modified MSC overrides the call forwarding functionality this time and makes a call to the mobile number of RBT subscriber. When RBT subscriber answers the call, both parties are then connected.

Technology requirement for RBT has three key aspects: (a) Telecom network switching system to support RBT (forward the call and route to RBT platform to connect to a custom tone), (b) RBT platform that accepts call from switch and connects it to the audio tune of user's choice, and (c) Public interface — IVR, where user can listen to range of audios and select a tune(s) of her/his choice as presented in Figure 7. Telecom operator managed the changes in core switch development, system integration, and billing-related activity. RBT platform development and application development was led by VAS provider independently. Network switching technology that supports the aspect (a) is facilitated by telecom operator, while RBT platform, contents (tunes, music), and application (user interface, subscription management, record for billing) are developed and provided by VAS provider. RBT solution for mobile subscriber is deployed within operator's service ownership as a managed service by mobile VAS provider. However, the entire system has been implemented and integrated with network elements by the operator.

Development and Implementation Stage

To illustrate the development process of RBT service, authors define service in terms of its functional characteristics (*Y*1, *Y*2, *Y*3, *Y*4, and *Y*5) as depicted in Figure 8. Two technological capabilities (TC(s)) — TC1 (basic communication network for voice call — GSM and CDMA) and TC2 (modified call routing capability in MSC) were identified as essential by the telecom operator to develop the services.

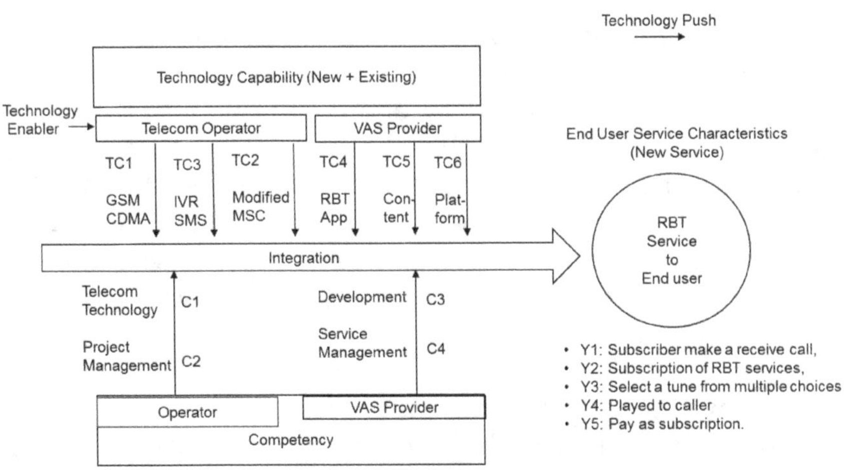

Figure 8. Process for RBT service innovation (Block "A" of Figure 6).

Also, telecom operators' existing capability TC3 — IVR and SMS capability were also considered for the service. Further, the capability of the mobile VAS provider — TC4 (RBT application for hosting of contents, custom selection of music, connection to a caller and then charging), TC5 (capability to host and manage content), and TC6 (the RBT platform to provide an ability to receive calls from switch to the RBT application for processing capability) were utilized for service development. TC(s) combined with competencies of telecom operator related to telecom technology (*C*1) and project management (*C*2), along with the VAS provider's competencies in application development (*C*3) and service management and operations (*C*4) led to the production of new service, RBT. The integration of [TC] and [*C*] provided by the telecom operator and VAS provider resulted in the desired functional characteristics of the new service [*Y*].

Case 2: Smart Energy Meter

Ideation Stage

One of the privately held Indian electric utility company observed that a small portion of consumers (20%–30%), industrial consumers, contribute about 70% of billing revenue. Delay in bill processing for such consumers or default in payment resulted in cash-flow problems and losses. Currently, meter reader personnel go from door to door, take the meter reading, and note it down manually, and then bring it to the utility administration office. With these requirements identified, the utility company conceptualized the project of introducing SEM that can be accessed remotely using existing technology and service, illustrating market-pull approach of the service innovation. The flow of the information in SEM service was ideated as shown in Figure 9.

(i) Energy utility company can access the energy meter (installed with subscriber identity module (SIM) card and communication modem) using the SMS functionality via

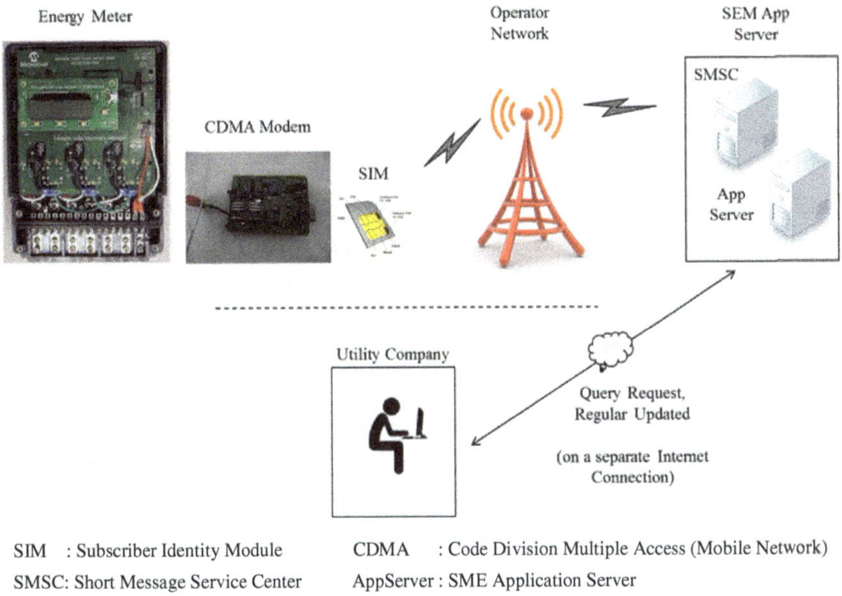

Figure 9. Information flow in SEM service.

connecting SEM application server over telecom operator network.

(ii) Utility company personnel can make query request to read the status of energy meter, for example, read energy consumption or also make changes to the configuration setting. For each request, a separate SMS is to be sent by SEM application sever to SIM (installed inside the communication modem).

(iii) This message is authorized and then interpreted for a predefined action (e.g., read, run routine program to connect supply, disconnect supply, etc.), and appropriate response is sent to SEM application server over SMS.

(iv) SMS is then received by SEM application server, which is sent to utility company's personnel as message response to the query made for a particular energy meter.

The utility company approached telecom operator "*X*" to work jointly to develop SEM service as a B2B service. Telecom operator took the lead role in facilitating and developing the solution. Telecom operator formed a cross-functional team comprising the domain representative and experts from telecom, IT, and marketing. Representatives from VAS provider (the utility company) were also part of the cross-functional team to finalize the solution architecture and design as per the market needs. Specialized tasks like SIM programming, meter integration application programming interface (API), and module development task, were carried out with the help of VAS provider.

Development and Implementation Stage
The development process of the SEM service required TC(s) of VAS provider and the telecom operator as presented in Figure 10. Two existing capabilities of telecom operators — TC1 (code division multiple access (CDMA) network and SMS), TC2 (CDMA modem capable of sending and receiving SMS), and one existing capability of VAS provider TC3 (meter system access) along with VAS provider's competencies are used by the telecom operator in creating the new capability — TC4 (meter system program), the system program at meter that performs control routine function (customized) and manages it by sending text messages (for interpreting SMS message as "command" and run predefined routine like "Read" message which results in getting the reading, transforms that reading, prepares text, and sends SMS with date and time stamp). A new capability TC5 — meter application management is also created at the VAS service provider end (the utility company) that incorporates the end-to-end process of VAS provider and associated IT applications (web access, utility's subscriber provisioning, billing, and access control). Telecom operator competencies in telecom technology (C1), project management (C2), system integration (C3), and IT application development (C4) combined with VAS provider competencies in electrical energy domain (C5) and field support operation (C6) were required for creating SEM service.

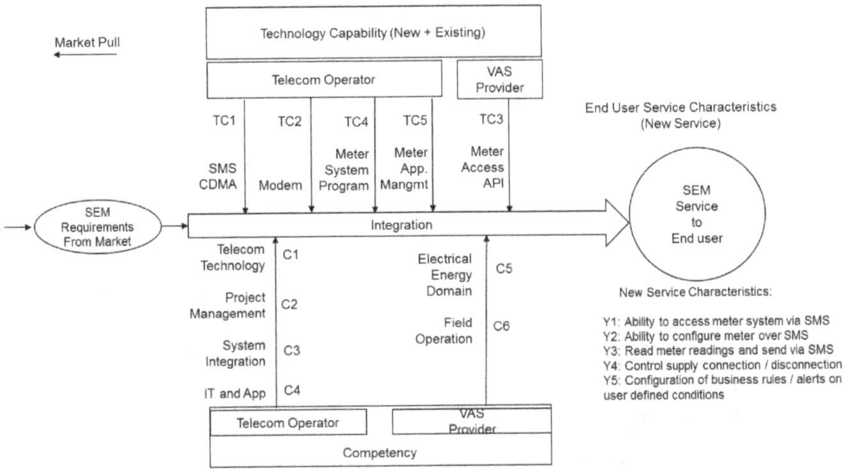

Figure 10. Process of SEM service innovation (Block "A" of Figure 6).

Case 3: WhatsApp Messenger

Ideation Stage

The engineering team at WhatsApp Inc. envisioned a multimedia instant messaging service — "WhatsApp" as a cross-platform service which uses internet data to send and receive messages — voice clips, text, images, videos, and location tags illustrating technology-push approach of service innovation. This service uses mobile data service for first-time registration; however, later operation can use any other internet access mode besides mobile data. WhatsApp services have been developed, deployed, and managed independently, like an over-the-top (OTT) service which is fast proliferating the market. OTT service uses telecom operator's infrastructure to reach to end user, in an almost similar way internet applications provider uses an Internet Service Provider (ISP). Integration of technological capabilities between the service provider and telecom operator is absent. Telecom operator does not own the end user for this service and all the service provisioning, access, and control are with VAS provider — WhatsApp, in this case, making telecom operator as a pass thru'.

Development and Implementation Stage

The development and implementation of WhatsApp are based on our analysis of information available in secondary literature [Reed, 2012, 2014; Umathe *et al.*, 2015]. The engineering team at WhatsApp Inc. (VAS provider) visualized a multimedia instant messaging service, with the set of functional requirements (*Y*) as presented in Figure 11. Different TC(s) were employed for service development and its implementation. TC(s) existing with open source community — TC1 (operating system (FreeBSD), web server (YAWS), database (Mnesia, Casandra), programming language (Erlang), and server scripting language (PHP)), TC2 (frameworks (Contus-Fly framework)), and protocols (customized instant messaging protocol) are utilized for the service development.

New TC(s) — TC3 (custom plugins, bridges for connecting different modules) and TC4 (WhatsApp client and server application for different operating systems like Android, iOS, Blackberry, Windows) are developed by VAS provider. In addition, TC5 (cloud computing capability) and TC6 (internet capability) of the cloud service provider (Amazon), and TC7 (mobile network), TC8 (SMS), and TC9 (data) of mobile operator

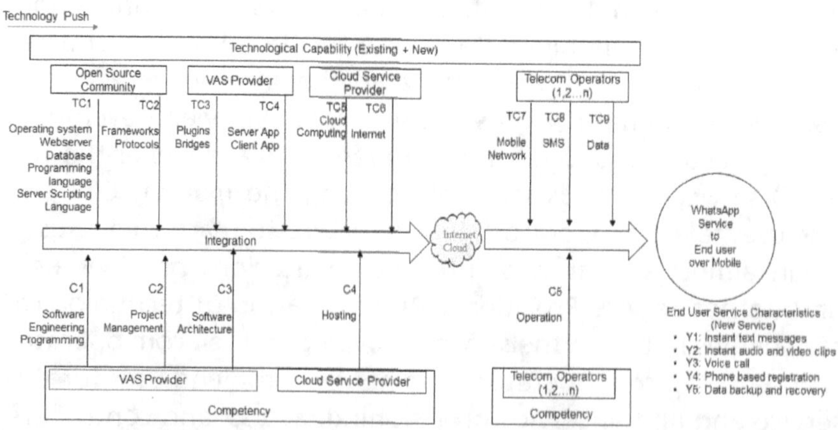

Figure 11. Process of WhatsApp service innovation (Block "A" of Figure 6).

are utilized to provide the service. Competencies of VAS provider — software engineering and programming (C1), project management and business acumen (C2), software architecture for massive scalability and operation (C3), along with the competency of cloud service provider to host the content and application (C4) and telecom service operational competency (C5) of telecom operator are used to develop and implement the service.

4.2. *Mapping of business processes on eTOM (Operations and Delivery)*

Case 1: RBT

Various process and subprocesses required to deliver the service, are analyzed upto eTOM level 3 and summarized to level 1, as depicted in Figure 12. The infrastructure required by RBT is deployed in premises of operator and, hence, an integrated eTOM is shown for operator enterprise. Telecom operator jointly discusses the features of RBT services and operational flow with mobile VAS provider. Operator takes the lead in core switching infrastructure, billing, CRM, while mobile VAS provider takes lead in RBT service development. Few custom interfaces have been developed to integrate additional network elements. Customers are owned by the operator, while VAS provider plays the role of in-premise managed service provider. Fulfillment and service assurance were jointly carried out. However, operations were as good as for any other telecom services. It was found that the business processes for the operation of RBT service ran smoothly and fit into the eTOM.

Case 2: SEM

Analysis of various processes and subprocesses to deliver the SEM service are analyzed upto the eTOM level 3 and summarized to the eTOM level 1. An integrated process mapping on eTOM is done in a similar way as in case 1, as shown in Figure 13. The telecom operator takes the lead in service, resource, and

◊ VAS Provider

• Telecom Operator

		Strategy Infrastructure and Product					
		Strategy		Infrastructure Life Cycle		Product Life Cycle	
		Telecom Operator	VAS Provider	Telecom Operator	VAS Provider	Telecom Operator	VAS Provider
Ringback Tone	Marketing and Offer Management	•	◊	•		•	◊
	Service Development and Management			•			◊
	Resource Development & Management			•			◊
	Supply Chain Development and Management			•		•	◊

		Operations					
		Fulfilment		Assurance		Billing	
		Telecom Operator	VAS Provider	Telecom Operator	VAS Provider	Telecom Operator	VAS Provider
Ringback Tone	Customer Relationship management	•	◊			•	
	Service Management & Operation	•	◊	•	◊		◊
	Resource Management & Operation	•	◊	•	◊		◊
	Supplier / Partner Relationship Management	•	◊	•	◊	•	

Figure 12. Integrated process mapping on eTOM (telecom operator and VAS provider) for RBT.

supply chain development and management, while marketing and offer management have been solely undertaken by VAS provider. The telecom operator and the VAS provider jointly decide on the product development life cycle, with the operator taking the lead in execution. Although fulfillment and service assurance were jointly carried out, system integration was complex and was not fully implemented at the system level.

Case 3: WhatsApp Messenger

WhatsApp application is positioned as a B2C service — developed, deployed, and managed independently by an application

◊ VAS Provider (An electricity utility company)

• Telecom Operator

		Strategy Infrastructure and Product					
		Strategy		Infrastrature Life Cycle		Product Life Cycle	
		Telecom Operator	VAS Provider	Telecom Operator	VAS Provider	Telecom Operator	VAS Provider
Smart Energy Meter	Marketing and Offer Management		◊	•		•	◊
	Service Development and Management	•		•		•	◊
	Resource Development & Management	•		•		•	◊
	Supply Chain Development and Management	•		•		•	◊

		Operations					
		Fulfilment		Assurance		Billing	
		Telecom Operator	VAS Provider	Telecom Operator	VAS Provider	Telecom Operator	VAS Provider
Smart Energy Meter	Customer Relationship management	•	◊		◊	•	
	Service Management & Operation	•	◊	•	◊	•	
	Resource Management & Operation	•	◊	•	◊	•	
	Supplier / Partner Relationship Management	•	◊	•	◊	•	◊

Figure 13. Integrated process mapping on eTOM (telecom operator and VAS provider) for SEM.

service provider (ASP) (WhatsApp Inc.). The telecom operator uses the eTOM for telecom operation and service management. Service framework for WhatsApp operations does not have any integration with any telecom operators, though it uses telecom operators' subscribers and network for service delivery. Mapping of business processes in the case is based on our analysis of the service characteristics and information available in secondary literature. The business process at WhatsApp provider is not known. It is assumed that the application provider has business processes similar to as in the eTOM. ASP has complete ownership of conceptualizing the service, market strategy,

◊ Application Service Provider (providing a VAS) • Telecom Operator		Telecom Operator				Application Service Provider (WhatsApp)		
		Strategy Infrastructure and Product				Strategy Infrastructure and Product		
		Strategy	Infrastructure Life Cycle	Product Life Cycle		Strategy	Infrastructure Life Cycle	Product Life Cycle
WhatsApp Services	Marketing and Offer Management				Geographically separate location Pass-thru' services so no system level integration	◊	◊	◊
	Service Development and Management					◊	◊	◊
	Resource Development & Management		•			◊	◊	◊
	Supply Chain Development and Management					◊	◊	◊
		Telecom Operator				Application Service Provider (WhatsApp)		
		Operations				Operations		
		Fulfilment	Assurance	Billing		Fulfilment	Assurance	Billing
WhatsApp Services	Customer Relationship management							
	Service Management & Operation					◊	◊	
	Resource Management & Operation	•	•			◊	◊	◊
	Supplier / Partner Relationship Management			•				◊

Figure 14. Integrated process mapping on eTOM (telecom operator and VAS provider) for WhatsApp.

developing, and implementing services. Telecom operator has no role for service-related development, resource development, and supply chain development and management. Application provider provides the entire range of operation life cycles processes "FAB" — provisioning, resources, fulfillment, subscription, and CRM.

In addition, the CRM aspect of WhatsApp helpdesk to directly interact with customers is lacking. Telecom operator's role is limited to providing a resource for handling additional traffic and maintaining an adequate QoS. A process mapping on eTOM was conducted in a similar way as in case 1 as shown in Figure 14.

5. Findings from Case Analysis

Findings from the analyses of cases as presented in Tables 2 and 3, respectively, indicate that each case represents a distinctive group of mobile VAS. It also represents evolutionary stages in mobile VAS management and end-to-end delivery perspectives.

Table 2. Findings from analysis on the process of service innovation.

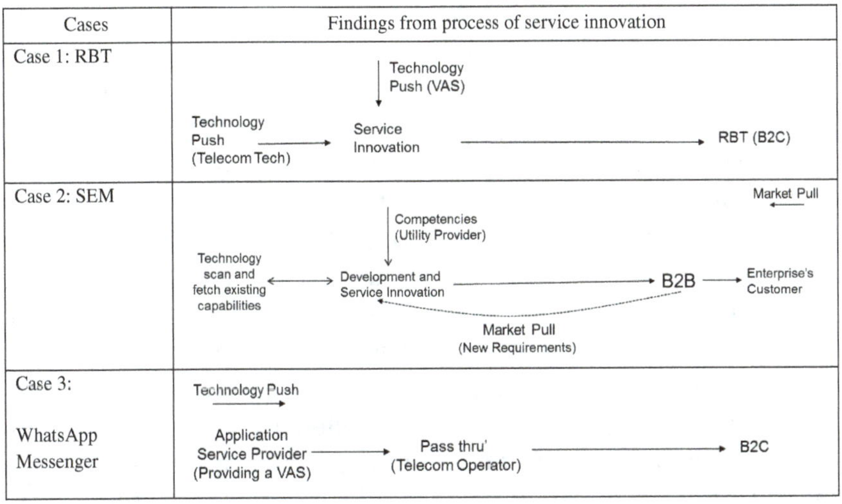

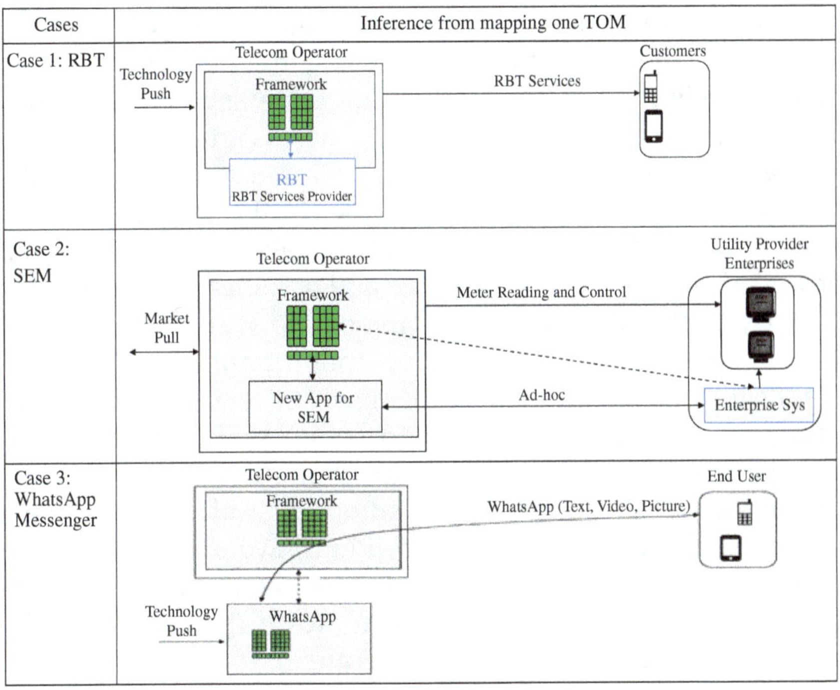

Figure 15. Inference from eTOM mapping analysis.
Note: *ad hoc* represents lack of process Standardization and proprietary implementation specific to operator.

The findings within Indian context are in line with the study of Eduardo and Sato [2014] indicating the adoption of platform-based approach by telecom operators for the purpose of reusability, customization, and derivative products around mobile VAS. Indian telecom operators have been adopting both technology-push and market-pull approaches for mobile VAS service innovations with unique service development process, for a unique set of requirements involving multiple partners. Case analysis indicates that RBT service was pushed by the operator leading to technology capability enhancement of telecom operator as shown in Table 2. Service development included significant innovation and competency brought by both the partners. However, the SEM service development was pulled by the market need (utility company's need) and involved operational competencies of both — telecom operator and VAS provider (utility company). Going beyond the conventional way to provide service presents an entrepreneur role of the telecom operator. Further, WhatsApp case represents an emerging trend of pushing the mobile application to the market and underlines the potential of OTT services as depicted in Table 2. There is little collaboration between ASP and telecom operator. Further, the analysis indicates that pull service involves the user in the service development process from the initial stage. This makes the development process complex, more interactive, and nonlinear. However, the service adoption by end users is immediate as compared to the "push" approach. Findings suggest that interplay among competencies and capabilities create new TCs and services. These newly created capabilities lead to future service innovations. The characteristics of new service innovations provide feedback for technological requirements leading to development of new TCs, and thus technological innovations. Evolution of services passes through simple in-house managed services (case 1), to an entrepreneur role of telecom operator (case 2), to an emerging trend of OTT as a pass-thru' service (case 3).

Further, Table 3 depicts inferences from the analyses of the three cases when mapped on eTOM. In the case of RBT, it was found that introduction of new service elements is brought into the scope of eTOM service management framework through custom interface development for seamless end-to-end service delivery. For in-premise managed service provided by the VAS provider, eTOM assimilates required changes, providing smooth and seamless extension of processes as in case 1. However, for enterprise services as in case 2, challenges are observed due to heterogeneity (processes, systems interfaces, and frameworks) and multiple ownerships of processes running between two enterprises. The case 2 reveals that business process is smooth at the SIP life cycle level but requires close integration of operations process (fulfillment, assurance, and CRM). Lack of standardization of operational processes resulted in high troubleshooting time and restricted extension of the service to other units/enterprises, for example, bill payment of utility service provider and bill payment of telecom operators, or disconnection of utility services and/or disconnection of telecom services. This is indicated as "*ad hoc*" in the diagram of Table 3. One major challenge that emerges from this case, is how eTOM will integrate with other process frameworks and facilitate collaborative efforts. In WhatsApp case, the process for SIP life cycle level is independent for both operator and application provider indicating the absence of collaboration. Operational and end-to-end delivery processes (FAB, CRM) require integration. Further, it was found that for OTT services (as in case 3), end-to-end service experience and subscriber ownership lacks clarity for end users. Telecom operators find it difficult to address the custom implementation of business process, bespoke contracts, pricing and billing, and finally in scaling up. How to manage end-to-end operations of mobile VAS with accountability and visibility is a question of concern.

Four key stages are observed in the mobile VAS development and delivery — Ideation, Development, Implementation, and Operations. As more and more aspects of the service delivery process cut across an enterprise's system, complexity increases, demanding close integration of development and operation processes. Subprocesses like — competency and capability mapping process, partner network architecture, and access interface need to be incorporated.

6. Discussions and Conclusions

Telecom ecosystem is an amalgamation of several sub-ecosystems with a number of players — supplier ecosystem, external ecosystems (such as content provider, content delivery network, creators, aggregators), and an IT ecosystem to manage these sub-ecosystems. This presents challenges such as significant degree of complexity, scalability, and cost. Geographical diversity, in dealing with enterprise requirements like localization and service level agreements (SLAs) pose further challenges. The analysis of the cases necessitates a need for a comprehensive framework that can incorporate challenges, addresses the multiple ownership issue, and facilitates collaborative development. Further, growing OTT services, evolving trends of technology convergence leading to service innovation in market place, require operators to redefine their roles and reposition themselves in the market place. It is foreseen that application service provider (ASP) and mobile VAS provider will further converge and possibly merge, absorbing the unique characteristics of each other. The findings suggest three evolving strategic positions in the market for telecom operators as: (a) Communication and computing Infrastructure Service Provider (IaaS), (b) ASP or as service broker, and (c) Solution provider — addressing the latent needs of society and creating domain-specific business solutions. Telecom operators may opt for one or more of these as their dominant roles based on their unique assets and strategies to reap the benefits. They can select to continue business by introducing mobile VAS in a managed services model as observed in RBT case. It is envisioned that telecom operators may increasingly adopt product-platform approach to optimize services and seek incremental revenue from its existing infrastructure, as observed in WhatsApp case. Also, telecom operator as IaaS will focus on providing and gaining revenue for communication infrastructure use and QoS. Mobile telecom operators can expand into Wireless ISP (WISP) and cloud-based infrastructure services, playing a dominant role as IaaS. In addition to IaaS, telecom operators may change their strategic position to become dominant mobile ASP providers. In this role, operator rejuvenates its services portfolio and creates additional revenue streams as observed in SEM case. The new role would require software and IT capabilities, coupled with their capabilities and experience in communication technology to reap the benefits of service innovations, particularly in B2B environment. This role of operators encompasses sector-crossing domain-specific solutions. Telecom operators have the ability to adopt emerging needs of enterprises in multiple domains and are therefore well-placed for sector-crossing service innovations across a range of sectors like health-care, banking, energy, agriculture, and education. This allows them to expand beyond communications and infrastructure services. This transformational role of orchestration among multiple ecosystems partners may well be called as "superCarrier".

The focus of this research was to understand the dynamics of technological and service innovation in the evolving context of technology convergence and its implication on role of operators in a digital economy. Three cases were studied and analyzed using: (i) Characteristic framework of service innovation and (ii) eTOM for telecom business processes. The research contributes to literature on mobile VAS in the Indian context. The framework for mobile VAS development is still in nascent stage and evolving with more empirical research on mobile VAS. Our study provides early evidence that mobile VAS development process has taken a significant shift from walled garden approach toward a modular approach, in which technological capabilities are treated as functional components to be integrated for the service innovation. The contribution of this research has been to emphasize the need for a framework for mobile VAS innovation, addressing current challenges and future requirements such as scale of service deployments across multiple enterprises and increased complexity.

Further, the research contributes to both theory and practice by highlighting the need to extend the eTOM framework for managing the emerging telecom service innovations (both B2B and B2C) involving multiple partners effectively. WhatsApp case has demonstrated a unique situation wherein innovation has taken place outside the realm of operator, making the VAS provider to connect directly with the subscriber, thus disrupting the linear value chain of mobile VAS [Li and Whalley, 2002; Pil and Holweg, 2006]. In linear value web, no matter which provider (VAS or telecom) fails to provide their share of service, it will affect the other partners including the end users. This causes ambiguity in subscriber ownership and managing billing relationship. At the same time, ensuring end-to-end service delivery is a challenge for the telecom operators, leading to services assurance issues. The role of telecom operator is then limited as a network operator eroding significant value which telecom operators enjoy today in managing subscriber relationship. If an operator wishes to work in an collaborative manner with such VAS providers, then existing telecom service delivery framework in its current status requires an extension to encompass service delivery framework used by VAS provider. This is how the end users' service experience can be managed by the operator effectively, and operator can take the ownership of end-to-end service experience of its own subscribers.

Our initial findings suggest that extending the capability of eTOM beyond the boundaries of telecom enterprise (to incorporate the heterogeneity of business processes and multiple ownerships across enterprises) to the external enterprise (B2B) is imperative. This would enable telecom operators to evaluate and manage their new service offerings optimally in the fierce competitive business environment. Further analysis would lead to a comprehensive framework for mobile VAS development encompassing subprocesses within each of core processes and interactions among them.

References

Abraham, R. (2007). Mobile phones and economic development: Evidence from the fishing industry in India. *Information Technology & International Development*, **4**(1):5–7.

Barras, R. (1986). Towards a theory of innovation in services. *Research Policy*, **15**(4):161–173.

BCG and IAMAI (2014). India@DIGITAL.BHARAT. Boston Consulting Group and IAMAI. Available at http://www.bcgindia.com/documents/file180687.pdf.

Berkowitz, S. (1997). Analyzing qualitative data. *User-Friendly Handbook for Mixed Method Evaluations*. National Science Foundation, Division of Research, Evaluation and Communication, USA, pp. 4–1.

Bigliardi, B., Ivo Dormio, A. and Galati, F. (2012). The adoption of open innovation within the telecommunication industry. *European Journal of Innovation Management*, **15**(1):27–54.

Birudavolu, S. and Nag, B. (2011). A study of open innovation in telecommunication services: A review of literature and trends. Available at SSRN: http://ssrn.com/abstract=2031639 or http://dx.doi.org/10.2139/ssrn.2031639.

Birudavolu, S. and Nag, B. (2015). The digital services landscape: An empirical study of open innovation ICT Forums. *International Journal of Management, IT and Engineering*, **5**(4):102–116.

Brown, S. and Eisenhardt, K. (1995). Product development: Past research, present findings and future directions. *Academy of Management Review*, **20**(2):343–378.

Caerteling, J. S., Halman, J. I. and Doree, A. G. (2008). Technology commercialization in road infrastructure: How government affects the variation and appropriability of technology. *Journal of Product Innovation Management*, **25**(2):143–161.

Chang, B. Y. (2011). Business process management of telecommunication companies: Fulfillment and operations support and readiness cases. *International Journal of Future Generation Communication and Networking*, **4**(3):73–86.

Coombs, R. and Miles, I. (2002). Innovation, measurement and services: The new problematique. *Innovation Systems in the Service Economy*, Ed. I. J. S. Metcalfe. Kulwer, Boston, pp. 85–103.

de Brentani, U. (1991). Success factors in developing new business services. *European Journal of Marketing*, **25**(2):33–59.

Deloitte and ASSOCHAM (2011). Mobile value added services (MVAS): A vehicle to usher in inclusive growth and bride the digital divide. https://www.deloitte.com/assets/Dcomindia/lolca%20 assets/Documents/Deloitte_aSSOChaM_MvaS_Study.pdf.

Denzin, N. K. and Lincon Y. S. (1994). *Handbook of Qualitative Research*, 2nd edn. Sage Publications, Inc.

Djellal, F. and Gallouj, F. (2001). Patterns of innovation organization in service firms: Portal survey resulted and theoretical models. *Science and Public Policy*, **28**(1):57–67.

Eduardo, C. and Sato, Y. (2014). Platform leadership of incumbent telecommunication operators: The case of BT 21st century network (BT21CN). *International Journal of Innovation Management*, **18**(2):1–37.

Frei, F. X. (2008). The four things a service business must get right. *Harvard Business Review*, **86**(4):70–80.

Gadrey, J. and Gallouj, F. (2002). *Productivity, Innovation and Knowledge in Services: New Economy and Socio-Economic*. Edward Elgar Publishing Company, Glos, UK.

Gadrey, J., Gallouj, F. and Weinstein, O. (1995). New modes of innovation: How services benefit industry. *International Journal of Service Industry Management*, **6**(3):4–16.

Gallouj, F. and Weinstein, O. (1997). Innovation in services. *Research Policy* **26**(4-5):537–556.

Glaser, B. S. and Strauss, A. (1967). *The Discovery of Grounded Theory: Strategies for Qualitative Research*. Weidenfeld and Nicolson, London.

Gupta, R. and Jain, K. (2012). Diffusion of mobile telephony in India: An empirical study. *Technology Forecasting & Social Change*, **79**(4):709–715.

Hacklin, F., Raurich, V. and Marxt, C. (2005). Implications of technological convergence on innovation trajectories: The case of ICT industry. *International Journal of Innovation and Technology Management*, **2**(03):313–330.

Helfat, C. E. and Peteraf, M. A. (2003). The dynamic resource-based view: Capability lifecycles. *Strategic Management Journal*, **24**(10): 997–1010.

Hertog, P. D., Wietze van der Aa and de Jong, M. W. (2010). Capabilities for managing service innovation: Towards a conceptual framework. *Journal of Service Management*, **21**(4):490–514.

IAMAI, IMRB, and I-Cube (2014). Internet in India. Business Research, Internet and Mobile Association of India and Indian Market Research Bureau, New Delhi, IAMAI-IMRB, http://www.iamai.in/rsh_pay.aspx?rid=4hjkHu7GsUU=.

ITU-T (2007). Interim view of an interpreter's guide for eTOM and ITIL practitioners, https://www.itu.int/rec/dologin_pub.asp?lang=e&id=T-REC-M.3050-200702-I!Sup1!PDF-E&type=items.

ITU (2014). ITU yearbook of statistics (2014), http://www.itu.int/en/ITU-D/Statistics/Pages/stat/default.aspx.

Kim, M., Song, J. and Triche, J. (2015). Toward an integrated framework for innovation in service: A resource-based view and dynamic capabilities approach. *Information Systems Frontier*, **17**:533–547, doi:10.1007/s10796-014-9505-6.

Korzachenko, O. and Getman, V. (2010). Improvement of business-activities in telecommunication enterprises by the eTM. https://ortus.rtu.lv/science/en/publications/8850/fulltext.

Lawson, B. and Samson, D. (2001). Developing innovation capability in organizations: A dynamic capabilities approach. *International Journal of Innovation Management*, **5**(3):377–400.

Li, F. and Whalley, J. (2002). Deconstruction of the telecommunications industry: From value chains to value networks. *Telecommunications Policy*, **26**(9):451–472.

Lipsman, A. (2014). Major mobile milestones: Apps now drive half of all time spent on digital, http://www.comscore.com/Insights/Blog/Major-Mobile-Milestones-in-May-Apps-Now-Drive-Half-of-All-Time-Spent-on-Digital.

Lovelock, C. and Gummesson, E. (2004). Whither services marketing? In search of a new paradigm and fresh perspectives. *Journal of service research*, **7**(1):20–41.

Menor, L. J., Tatikonda, M. V. and Sampson, S. E. (2002). New service development: Areas for exploitation and exploration. *Journal of Operations Management*, **20**(2):135–157.

Miles, M. B. and Huberman, A. M. (1994). *Qualitative Data Analysis: An Expanded Handbook of Qualitative Research*. Sage Publications, Inc., Thousand Oaks, CA, pp. 428–444.

Moller, K., Rajala, R. and Westerlund, M. (2008). Service innovation myopia? A new recipe for client-provider value creation. *California Management Review*, **50**(3):31–48.

Nijssen, E. J., Hillebrand, B., Vermeulen, P. A. and Kemp, R. G. (2006). Exploring product and service innovation similarities and differences. *International Journal of Marketing Research*, **23**:241–251.

Nesse, P. J. (2009). Open service innovation in telecom industry: A case study of partnership models enabling 3'rd party development of novel mobile services. Telenor Research & Innovation, Service Innovation Group.

Olla, P. and Patel, N. V. (2002). A value chain model for mobile data service providers. *Telecommunications Policy*, **26**(9):551–571.

Papastathoulou, P. and Hultink, E. J. (2012). New service development: An analysis of 27 years research. *Journal of Product Innovation Management*, **29**(5):705–714.

Payne, A. and Frow, P. (2005). A strategic framework for customer relationship management. *Journal of Marketing*, **69**(4):167–176.

Peppard, J. and Rylander, A. (2006). From value chain to value network: An insight for mobile operators. *European Management Journal*, **24**(2):128–141.

Pil, F. K. and Holweg, M. (2006). Evolving from value chain to value grid. *MIT Sloan Management Review*, **47**(4):72.

Reed, R. (2014). Scaling to millions of simultaneous connections. *Earlang Factory Conference 2012*. Available at https://www.youtube.com/watch?v=wDk6l3tPBuw.

Reed, R. (2014). That's billion with B: Scaling to the next level at WhatsApp Inc. *Earlang Factory Conference 2014*. Available at https://www.youtube.com/watch?v=c12cYAUTXXs.

Rothwell, R. (1994). Towards the fifth-generation of innovation process. *International Marketing Review*, **11**(1):7–31.

Saviotti, P. P. and Metcalfe, J. S. (1984). A theoretical approach to the construction of technological output indicators. *Research Policy*, **13**(3):141–151.

Shirahada, K., Belal, H. M. and Takahashi, N. (2015). Development of technology and service thinking for technical personnel: Action research at a Japanese monitor maker. *Technology in Society*, **43**: 191–198.

Song, L. Z., Song, M. and Benedetto, A. D. (2009). A staged service innovation model. *Decision Science*, **30**(3):571–599.

Tanner, J. C. (2010). Twenty years and changes. Telecom Asia, pp. 19–25.

Teece, D., Pisano, G. and Shuen, A. (1997). Dynamic capabilities and strategic management. *Strategic Management Journal*, **18**(7): 509–533.

Tether, B. and Howells, J. (2007). Changing understanding of innovation of services: From technological adoption to complex complementary changes to technologies, scales and organization. DTI Occasional Paper No. 9, http://www.cambridgeservicealliance.org/uploads/downloadfiles/DTIpaper9.pdf.

TM forum and ITU-T M.3050, (2007). https://www.itu.int/rec/dologin_pub.asp?lang=e&id=T-REC-M.3050-200702-I!Sup1!PDFE&type=items.

TRAI (2014). Highlights of telecom subscription data as on 30th November 2014. Telecom Regulatory Authority of India (TRAI), Press Release No. 04/2015.

Ulaga, W. and Reinartz, W. J. (2011). Hybrid offerings: How manufacturing firms combine goods and services successfully. *Journal of Marketing*, **75**(6):5–23.

Umathe, N., Tadvi, J. and Chauhan, L. (2015). Technology behind WhatsApp messenger. Available at https://www.youtube.com/watch?v=iyDXPRoEK2I.

Vargo, S. and Lusch, R. F. (2004). The four service marketing myths; Remnants of good-based, manufacturing model. *Journal of Service Research*, **6**(4):324–335.

Vargo, S. L., Maglio, P. P. and Akaka, M. A. (2008). On value and value co-creation: A service systems and service logic perspective. *European Management Journal*, **26**(3):145–152.

Wong, D. (2009). Adapting eTOM model to improve multimedia network management: A case study with a Taiwan multimedia service provider. *Fifth International Conference on Intelligent Information Hiding and Multimedia Signal Processing IEEE*, pp. 579–582.

Yin, R. (1994). *Case Study Research: Design and Methods*. Sage Publications, Beverly Hills.

Yin, R. K. (2013). *Case Study Research: Design and Methods*. Sage Publications.

Chapter 5

International Perspectives: Case of Europe

Chia-Liang Hung and Shu-Yu Yeh

1. Introduction

On February 22, 1999, the Japanese mobile operator, NTT DoCoMo, launched its pioneering mobile commerce system, i-mode. Subscribership increased rapidly, to 200,000 after the first 20 days, approaching one million after six months. User numbers continued to mushroom, reaching 3.8 million in 2003, roughly 80% of worldwide mobile Internet subscribers at the time [Funk, 2001]. By December 2005, i-mode had reached 45 million subscribers. As i-mode spread into Europe, it developed an image similar to that of other successful Japanese products, inevitably becoming a target of scholarly research.[a]

[a] i-mode services have already been launched in 14 countries outside Japan (to December 2005). The internationalization began with E-Plus of Germany in March 2002, followed by KPN Mobile in the Netherlands, KG telecom in Taiwan (now acquired by Far EasTone telecom), BASE in Belgium, and France's Bouygues Telecom in the same year. Spain's Telefonica Moviles introduced i-mode in June of 2003, as did Wind of Italy in November. Then, in June 2004, Greece's COSMOTE unveiled i-mode in time for the Olympics, and Telstra in Australia also premiered i-mode in November. Israel's Cellcom Israel and Russia's Mobile TeleSystems have launched the i-mode service in September 2005, followed by O2 in the United Kingdom and Ireland in October. Singapore's StarHub also launched i-mode in November. For information about i-mode international diffusion, please refer to http://www.nttdocomo.com/corebiz/imode/global/index.html; "NTT's global reach", *Wireless Week*, Vol. 6, Iss. 20, May 15, 2000; "Japan's i-mode becomes international mode," Global Telecoms

Research studies articulated by both internal developers at DoCoMo and external researchers have argued that the success of i-mode was driven not only by NTT's technological advances, but also the redefinition of traditional business models that i-mode offered [Beck and Wade, 2003; Funk, 2001; Kalakota and Robinson, 2001; Matsunaga, 2000; Natsuno, 2003; Ratliff, 2002]. By contrast, the collectively agreed-upon wireless application protocol WAP is unexpectedly laggard. However, the global diffusion rate of i-mode seemed to be much slower than the domestic rate, just 100,000 after the first six months, less than one-fourth of the projected 500,000.[b] Moreover, as Table 1 shows, after two years of diffusion, overseas subscribers' number just 2 million, one-tenth of Japan's domestic market of 20 million.

What has slowed adoption of i-mode overseas? Gawer and Cusumano [2002] have argued that NTT DoCoMo's local i-mode standard is not adequate for a worldwide standard because it has shortcomings of compatibility and expansion possibility. Examining the issue from another perspective, our research will explore institutional differences between Japan and overseas markets.

This chapter is organized as follows: Section 2 offers a conceptual framework for analyzing the mobile service industry using an institutional perspective. Section 3 will compare institutional differences between the Japanese domestic environment and overseas markets. Section 4 contains the discussion, including recommendations and implications for industry policy. The final section presents our conclusions and suggestions for further research.

Business, London, May/June 2002; "DoCoMo, Bouygues Agree on i-mode License in France," Jiji Press English News Service, Tokyo, April 17th, 2002. and the website of NTT DoCoMo, http://www.nttdocomo.com/corebiz/global/imode/index.html.

[b]In the beginning, most i-mode adopters in Europe were users of Germany's E-plus, with subscribers of Netherlands KPN in the number two slot. For details, see "KPN to delay i-Mode in Belgium," Europemedia, June 14, 2002, and "Take up of KPN Mobile's i-mode slower than expected," Europemedia, Amsterdam, August 13, 2002.

Table 1. i-mode subscribers, Japan and overseas, two years after introduction.

Date	Domestic subscribers (millions)	Date	Overseas subscribers*
1999.02	0	N.A.	N.A.
2000.02	4.4	N.A.	N.A.
2000.08	10.0	N.A.	N.A.
2001.02	**20.0**	N.A.	N.A.
2001.12	30.0	N.A.	N.A.
2002.12	36.0	2002.03	0.
2003.08	40.0	2003.05	0.5
2004.02	41.0	2004.02	**2.0**
2004.07	42.0	2004.07	3.0
2005.06	44.0	2005.06	5.0

Note: *Subscriber figures published by 5 earlier overseas operators, Taiwan's KG Telecom, Germany's E-plus, the Netherlands' KPN, Belgium's BASE, and France's Bouygres, and press releases from DoCoMo's press center.
Source: http://www.nttdocomo.com/presscenter.

2. Literature Review and Industry Analysis

There are two purposeful diffusion properties used in strategic marketing of information and communication technologies: the lock-in effect, which retains subscribers, and the network effect, which acquires more subscribers. Our discussion will explore the institutional transformation of telecommunications environments in the post-2G era, which may hinder amplification of these diffusion properties by telecom firms.

2.1. *Lock-in effect*

The first technology diffusion property desired by telecom firms is the *lock-in* effect caused by the switching cost, which increases continuously once the subscribers become used to a particular product [Arthur, 1988; Shapiro and Varian, 1998]. As long as the lock-in effect is strong, the firm will retain customers and sustain its market power. It is not difficult to find marketing strategies based on this effect including loyalty programs, long-term contracts, intensive education, and the technological specificity, all of which give incentive to customers to remain with their current firm by increasing switching costs. In the case of the 2G and 2.5G eras, more and more mobile subscribers are locked into long-term services by the subscription fees and subsidized mobile phones. Further, non-portable phone numbers and reusable SIM cards, which accumulate personal and business interconnections, lock subscribers into particular operators [Paavilainen, 2001]. However, these lock in strategies pose problems for operators when they want to launch new mobile phone standards, since consumers incur switching costs when changing to the new standards.

2.2. *Network effect*

The second important property of technology diffusion is the *network effect*, or network externalities, which are purposefully manipulated in order to increase the consumption utility of subscribers as the number of subscription increases [Shapiro and Varian, 1998]. In the telecommunications industry, the utility of connection usage increases as a square of the number of connecting nodes/subscribers, in accordance with Metcalfe's Law. This implies that market share is vitally important. Consequently, some IT firms have applied penetration pricing or a loosely appropriable IPR strategy to quickly accumulate sales volume, especially for newly-launched products, in order to increase the network effects. Products such as Microsoft's DOS operating system and the IE browser used this strategy [Kotler *et al.*, 1996; Shapiro and Varian, 1998; Yoffie and Cusumano, 1999]. In addition, some firms reinforced network externalities by using a complementary strategy of licensing or acquiring aggressively complementary assets [Gawer and Cusumano, 2002; Moore, 1995; Teece, 1987]. For instance, Nintendo invited many game vendors to develop video games in order to increase the utility of the Nintendo system [Shapiro and Varian, 1998]. Sometimes, if a leverageable installed base for product migration and the time-to-market benefit appears, innovators will cross the diffusion chasm immediately [Rogers, 1995; Moore, 1995]. For example, the Internet penetrated on the extant installed PC base and open protocols, TCP/IP, and HTTP [Yoffie and Cusumano, 1999]. In spite of the soft dimension of installed bases, the intangible enterprise reputation, and the bandwagon effect, they represent credible commitments to encourage collective adoption in the face of technological uncertainties [Rohlfs, 2001].

The IT and telecom firms strategically manipulate both of these technological properties, but outside institutional environments affect the success of these strategies. Hence, the transformation of the mobile institutional environment should be investigated prior to any analysis of new mobile technology diffusion.

2.3. *Characteristics of mobile institutional environments*

The mobile industry is more or less a regulated market, though deregulation began in the 1980s all over the world, especially after the emergence of the 2G era. The increasing number of operators, from three to five per market, facilitates market efficiency and consumer welfare even though there are not enough firms for perfect competition [Funk, 2002]. In fact, constraints deriving from institutional rules and organizational structures rather than the technologies themselves may influence the evolution of technology, though individuals or firms may intend to follow or create a particular path [North, 1990; Oliver, 1997]. The following discusses the structure of the mobile industry from the perspective of institutional economics.

(1) Vertical/Horizontal disintegration.

According to the value layer model, the multimedia Internet industry is composed of the content layer, the packing layer, the terminal device layer, the manipulation layer, and the transmission layer [Collis *et al.*, 1997]. Several companies can participate in a given layer with open interfaces only if they can offer valuable services as a whole. This implies a situation in which several firms compete for a limited number of rents. Connections in the vertically disintegrated mobile industry are sustained on one side by the content-related services and applications, which have a vast potential and have attracted many entrants, and on the other side by the network infrastructure and devices, which require investment capital for broadband enhancement [Maitland *et al.*, 2002; Sabat, 2002].

Thus, the mobile industry has been evolving toward a more specialized and more disintegrated institutional structure even as the interfaces between these structures have consolidated and matured.[c] The more standardized the interface is, the more participants will invest, and the more institutionally embedded the industrial clusters become [Abernathy and Utterback, 1978]. It is also difficult for a single firm to dominate the business and change the operating interfaces without coordination with players of other value layers. As a result of path-dependence, incumbents find it difficult to advance, since they are locked in by past investments and lack the market control to coordinate the collective switching cost for the whole value chain [Christensen, 1997]. Moreover, because of telecommunication liberalization, new firms have entered, increasing competition, and steadily diluting not only the monopoly power of any single firm, especially in the contents and terminal layers,

[c]As the mention of Steinbock [2003], the traditional telecommunication was served by an integrated mode, that the whole services from the installation of hardware infrastructure, the signal software, and service development were coordinated by the operator, such as the AT&T. But after the telecom deregulation, several firms focused on infrastructures like Ericsson, Lucent, Siemens, and Nokia, while some specialized in the manipulation software or browser like the Symbian, Microsoft, and Qualcomm, and moreover, many operators in the world purchased those above hardwares and softwares to provide connection services like Vodafone, DoCoMo, China Mobile, Cingular, and so on.

but also the network effect of the subscribers [Funk, 2002; Lera, 2000; Steinbock, 2003]. Therefore, the market of each value layer has been horizontally disintegrated and is now served by several alternative suppliers. This means that the suppliers in each value layer must aggressively differentiate. Under such conditions, existing open standards sometimes may result in market balkanization, as with the UNIX OS [Yoffie and Cusumano, 1999]. Consequently, the dispersion of individual innovation has discouraged coordination among firms and led to the side effects of incompatibility in the horizontally disintegrated market structure.[d]

In such a situation of disintegrated but multiple complements, especially in the IT and communication industry, the firm possessing *leadership* and superior technological *coordination capability* within the whole value chain could champion the collective adoption of new standards or business models in ways that government cannot [Baba *et al.*, 1996]. This process resembles the informal but flexible approach for adaptation under the formal industry structure in terms of institutional economies. In addition, the population *scale* of potential subscribers may also give the firms incentive to innovate services with the benefit of economies of scale, especially for the mobile industry, in which there are often high sunk costs [Funk, 2001].

---

[d]In the open standard telecom context, most service operators and device vendors intended to differentiate for distinctiveness, so they had mutually allied as groups for complements and launched new mobile services based on different new standards or the existing standard derivatives. For examples, NTT DoCoMo adopted the derivative cHTML and the special i-mode protocol and collaborated with Japanese content providers and the phone vendors. The American operator Verizon also cooperated with 35 content providers, phone vendors, and software developers for its advanced mobile services on the evolutionary standards of WAP, Brew, EDGE, and MMS, while the Sprint PCS also allied with 100 complementary firms. Therefore, the groups compete mutually [Gawer and Cusumano, 2002].

(2) The decreasing degree of market concentration.

Due to the privatization, liberalization, and horizontal disintegration of telecommunications, the mobile market has fragmented into three to five operators or more, in contrast to the past, where one firm maintained a dominant position [Funk, 2002; Laffont and Tirole, 1999; Steinbock, 2003]. The increasing number of players will lead to a decreasing degree of market concentration [Scherer, 1990]. As a result, the lower degree of market concentration inevitably decreases the economies of scale to the service providers as well as the network externalities to the subscribers. If one operator wants to innovate, it has to consider the technology choices of other firms and come up with ways to counteract the bandwagon effects of extant technology promotion by its competitors. Though innovative operators will invest strategically to differentiate themselves from other firms, they may be not only locked into a particular technology because of sunk costs, but also locked out by oligopoly substitution, due to errors in technology choices. So, in many cases even would-be innovators will wait and see how things go [Laffont and Tirole, 1999]. Widespread hesitation among firms may thus slow the progress of technological advancement in the industry.

The *initial state* of market and the launch timing may affect the degree of competition and the speed of market penetration. If the innovators purposefully choose an unexploited market space even through lucky timing, they will target potential new buyers by bypassing the cost of persuading existing subscribers who may been locked in by past durable purchasing [Beck and Wade, 2003].[e] So, the market of *lower*

[e]Unique features of the Japanese market favored the development of mobile internet services in Japan. Japan has long been behind the U.S. in PC and Internet penetration, largely because most Japanese people do not have room for a desktop in their homes and lack familiarity with the keyboard. But millions of Japanese have grown up playing video and pocket computer games — the so-called "push-button" generation. For them, the handset is

penetration rate locked by the earlier technology will be preferred, where the later majority will enjoy the second-mover advantage contributed by the later but newer technologies.

(3) Unstable evolutionary infrastructure.

Mobile phones lack add-on attributes, being essentially similar to the network card plugged into the PC slot so as to access the Internet. Hence, mobile subscribers have to abandon the original phone devices and repurchase new ones suited for new mobile Internet services. Therefore, mobile operators have to patiently promote advanced mobile services to entice consumers to overcome the hesitation caused by the costs of switching a mobile phone during its lifecycle, which is usually more than one year but shorter than the lifetime of a PC [Funk, 2001; Kalakota and Robinson, 2001; Natsuno, 2003; Paavilainen, 2001].

However, the migration path to mobile Internet is more difficult than that of PC to the fixed-line Internet. After reviewing the evolution history of the mobile mainstream standard GSM, we have found several network infrastructures, including HSCSD, GPRS, EDGE, and even UMTS of 3G.[f] Which one is the best bet? The 2.5G GPRS has emerged as the dominant mobile Internet network after a long period of hesitation. There has also been hesitation among operators in choosing between the

---

a familiar alternative to the computer keyboard. Nokia and others are also encouraging operators to do things on small devices. Consequently, Japanese access the Internet through mobile phones, moving beyond the desktop Internet. For details, see Natsuno [2003] and also refer to the industry reports: "Amazing DoCoMo" and "It Could Obviate the Need for Cable" in BusinessWeek on January 17, 2000, which are available at http://www.businessweek.com/archives/2000/b3664010.arc.htm.

[f]In addition to the mainstream standard GSM, CDMA is second in the market, covering most of Korea, Japan, US, Hong Kong, and China. Its evolutionary paths are cdmaOne (IS-95A & IS-95B) of 2G and cdma2000 (1x, 1x EVDO, &1x EVDV) of 3G. Most of the South Korean CDMA development has used the WAP protocol. See Funk [2002] and Steinbock [2003].

open WAP, the popular MMS, Java, or even the proprietary i-mode for the application layer standard. The large number of choices has perplexed both operators and mobile equipment manufacturers, who also intend to compete for dominant positions by locking in subscribers. Many alliances compete mutually to maintain preferred protocols, resulting in an unstable and unclear market evolution. With the evolutionary path of the infrastructure still unclear, operators have hesitated in adopting new technology and delayed the launch of new mobile Internet businesses. For example, NTT DoCoMo was a key member of open agreed standard WAP forum, but it developed the other proprietary i-mode for its own uses. Such a big industrial player that participated in the commonly agreed technology development and protocol definition could give up the open standard unexpectedly, or to promote the bandwagon effect for persuading many smaller followers. Therefore, the hesitation of migration investment resulted in the delay of diffusion.

In this unstable situation, the coordination capability and the reputation of firms accumulated in past technological experience may be used as *leverageable installed bases* that would facilitate successful technological migration.[9] This is an informal but credible approach under the formal industry structure. Once the collective investments have been made and synergies

[9]In tracing back to the telecommunications industry, equipment vendors and telecom operators have cooperated intensively, especially in the national-owned monopoly era or in the PTT (Ministry of Post, Telecommunication, and Telegraph) stage. This special *ventel* relationship collectively drove the technological evolution just as *wintel* did the PC industry. For example, the *ventel* relationships sprung up between the equipment vendors Ericsson and Nokia and the telecom operators Telia and Sonera in the Nordic area, as well as between Alcatel and France Telecom, Siemens and Deutsche Telekom, Motorola and AT&T, and NTT and the traditional Den Den family (NEC, Fujitsu, Hitachi, Oki). So, if the technological giants and operators cooperate to initiate the advance of technology in a vertically and horizontally disintegrated environment, the bandwagon effect will take off more easily and rapidly. See Steinbock [2003] for the details.

between complementary assets are properly exploited, the great bandwagon effect of new technology would eliminate the problem of locked-in assets and hesitation on new technology, in turn leading to exploitation of economies of scale that will bring about network effects [Baba *et al.*, 1996; Natsuno, 2003; Steinbock, 2003].

2.4. *Conceptual framework of research*

From the foregoing discussions of institutional transformation and candidate resolutions, we have developed a research framework for mobile industry analysis (Figure 1). The institutional transformations of 2G, including the vertical/horizontal disintegration, the decreasing market concentration, and the unstable evolutionary infrastructure in the middle, will negatively impact the performance of new standards. However, certain conditions such as informal rules of institutional economics may be applied to mitigate the negative institutional impacts on the diffusion performance caused by telecom industry transformations.[h] In Figure 1, the total effect leveraging from the strategic institutional levers in the left end to the performance indicators in the right end is positive. In other words, if more applicable levers appear, more possibilities to increase the diffusion performance even under the adverse institutional environment of telecommunication will exist.

The mobile industry, including the equipment, operators, content vendors, and application vendors, needs effective strategies to develop positive feedback loops that reinforce

[h]According to the North's institutional economics, some pervasiveness of information rules arose from extensions, elaborations, and modifications of formal rules, while some were caused from the social sanctioned norms of behaviors and internally enforced standard of conduct [North, 1990]. So the special path dependence that learned from continuous adjustments to the governmental regulation, the trust among the particular clustered firms, and the relationship and the reputation between sellers and buyers may be the facilitator for industry transformation than other situations.

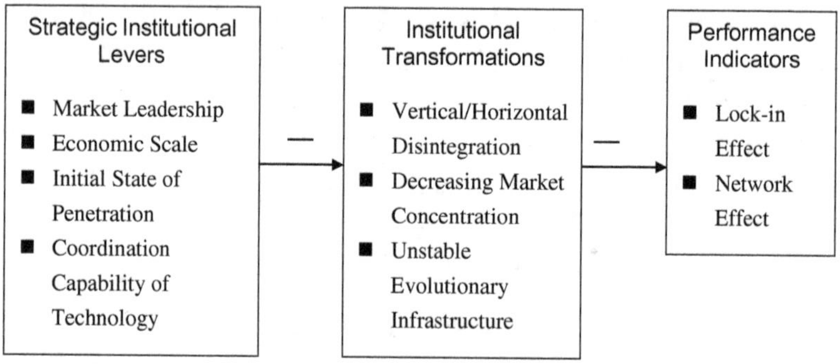

Figure 1. The research framework.

diffusion of mobile Internet services under the current conditions of environmental hindrance. The operator in the gateway position of the mobile Internet is especially well-situated for the initiative activities [Funk, 2001]. In the next section, we will compare the institutional differences in leveraging the negative impacts caused by environments disadvantageous for 2G. Even though NTT DoCoMo and its overseas licensed operators took responsibility for the diffusion of the i-mode mobile Internet, the performances actually differed. We will also identify institutional situations in overseas markets that were unfavorable to i-mode, and compare them to the Japanese domestic market.

3. Institutional Comparisons of International i-mode

Several research studies have already offered critical insight into the successful diffusion of i-mode from the marketing, the social, and the technological points of view. In this paper, we will explore the diffusion of i-mode into overseas markets through an institutional comparison of Japan's domestic market and overseas markets [Beck and Wade, 2003; Funk, 2002; Ratliff, 2002; Vacca, 2001].

3.1. *DoCoMo and innovation*

DoCoMo, whose major investor is the government-owned telecomm operator NTT, controls over half of the 100 million Japanese mobile subscribers. Even though DoCoMo implemented the NTT-specific 2G system, the Personal Digital Cellular (PDC) wireless standard that is different from the global choice of GSM, DoCoMo has developed a suitable mobile business model and pioneered mobile Internet services before most western countries. Meanwhile, it remains ahead of its two largest competitors in the Japan market, J-phone (acquired by UK's Vodafone now) and KDDI [Funk, 2001]. DoCoMo's approach to the conquest of the institutional difficulties of creating mobile Internet was quite innovative.

First, on the *infrastructure* side, DoCoMo immediately consolidated a packet-based PDC-P network, which is suitable for mobile data transmission with an average rate of 9.6 kbps, in order to *pave the technological migration route*. This move mitigated the investment hesitation caused by the apparently limitless upgrade path that is usually taken in other GSM areas.

Moreover, DoCoMo proposed several attractive *promotion* programs to induce mobile subscribers to transmit data. One of them involved applying a *subsidization* strategy to mobile phone manufacturing vendors in order to launch the inexpensive new cellular phones, so as to spread i-mode quickly. On average, DoCoMo subsidized phone manufacturers at a rate of US$300 per phone while the subscribers of i-mode paid for the new Internet phones at a *penetration* price of roughly US$50. This strategy differed from the skimming pricing strategy traditionally used in selling new technology products [Funk, 2001; Kotler *et al.*, 1996]. As a result, the phone manufacturers never hung back on mass production of a new product model, and existing subscribers were willing to switch to i-mode. More importantly, DoCoMo *aggressively licensed* out i-mode models to many vendors such as Mitsubishi, Fujitsu, NEC, Panasonic, Sony, and Sharp, in order to obtain the collective bandwagon effects of suppliers as well as to stimulate the competition effects that would benefit subscribers.

Additionally, DoCoMo also actively invited the content providers to develop many profitable and useful complements by *leveraging the installed base*, i.e., cHTML, which is a compact version of Internet markup language adapted from the familiar HTML. For the complementary utilities to the subscribers and providers, DoCoMo also established advertisement companies to support the content providers for packaging the mobile i-mode pages [Natsuno, 2003; Ratliff, 2002].[i] Moreover, DoCoMo motivated the subscribers to remain online for receiving information from the mobile websites anytime by means of an asymmetric pricing strategy, which is an *efficient incentive* of receiving being cheaper than sending, in order to encourage the subscribers' downloading amounts. As a result, all of these strategic maneuverings together helped Japan's mobile Internet commerce to take off.

In addition to the purposeful strategies adopted by DoCoMo, the special institutional facilitators of the Japanese industry structure and DoCoMo's path-dependent capabilities also require elucidation.

3.2. *Comparisons of institutional levers*

First, the *market leadership* may matter. Even though the new entrants KDD, IDO, and DDI have been competitive after deregulation, DoCoMo is still the mobile leader, controlling over half the mobile market in Japan. By contrast, the first five

[i]NTT DoCoMo established DoCoMo.com (D2) to give expert advice on content production to interested service providers and, sometimes, even to invest in their businesses. D2 communications is also a facilitator, designing a readable i-mode-sized advertising banner for i-mode complements, and D2 serves as a mobile advertising mediator and packager by bringing the contents of TV, FM, newspaper, and magazine into the mobile network and phone. For details, see Natsuno [2003] and Vacca [2001].

overseas i-mode licensed operators occupied no more than 50% of market share in their countries, just 21% on a weighted average. Further, overseas i-mode licensees are falling behind the leader in their respective markets, occupying the third or fourth position, with the exception of KPN mobile. If a market leader has the majority of subscribers, possesses advanced technologies, and implements effective marketing strategies, it will enjoy lower switching costs for new services than its smaller rivals. Table 2 shows a comparison of the mobile industry structures in Japan and abroad, including the i-mode licensed operator, the respective launching time, the market share and position, the size of the installed base, the national mobile penetration rate, key supporting hand device manufacturers and content providers, and the standard of network infrastructure.

Second, the *economic scale* of subscribers may matter. There are 120 million Japanese, a market scale great enough to attract many handheld device manufacturers and mobile content vendors. Even though DoCoMo adopted the NTT-specific infrastructure and its own proprietary i-mode protocol rather than the open WAP, the Japanese market had a scale great enough to sustain localized technologies. Typically, the profitable volume of a new mobile phone model to the manufacturer is at least 500,000 units, generally ordered in advance by a given telecom operator [Funk, 2001]. This scale is a difficult challenge for operators located in countries with small populations, even in the mobile-advanced Nordic countries. Moreover, great economic scale attracts relatively more suppliers to target the same market, leading to product proliferation and diversification, diffusing new mobile technology more rapidly, and speeding transformation of the market. However, the overseas i-mode operators, confronting smaller market scale, are at a disadvantage in attracting complements and are often in competition with the WAP or similar services of the local leaders. Even though the aggregate population of western Europe is comparable to that of Japan and diffusion

Table 2. The institutional comparisons among i-mode licensed areas.

Country	Japan	Germany	The Netherlands	Taiwan	Belgium	France
Cooperating operator	NTT DoCoMo	E-plus	KPN Mobile	KG Telecom	BASE	Bouygres Telecom
i-mode service launched	1999.2.22	2002.3.16	2002.4.18	2002.6.20	2002.10.15	2002.11.15
Size of installed base (millions of subscribers)	25 million	7.5 million	5.2 million	3.7 million	1 million	6.0 million
Market share/ position	60%/1st	12%/3rd	42%/1st	16%/4th	14%/3rd	19%/3rd
National-wide penetration rate	30%	73%	75%	106%	79%	65%
Cooperating device manufacturers	Mitsubishi, Fujitsu, NEC, Panasonic, Sony, Sharp, etc.	NEC, Toshiba	NEC, Toshiba	NEC, Toshiba	NEC, Toshiba	NEC, Toshiba
Network standards	PDC/PDC-P	GSM/GPRS	GSM/GPRS	GSM/GPRS	GSM/GPRS	GSM/GPRS
Official menu sites	67‡	400	60	130	60	90

Note: ‡Total launched i-mode sites exceeded 550, including many independent content providers connected by the i-search mobile search engine.
Source: http://www.nttdocomo.com/corebiz/imode/global/index.html and http://www.itu.int/ITU-D/ict/dai/high.html.

was actively coordinated by DoCoMo, locality properties such as language, usually emphasized by mobile services, may instead lessen the aggregation effect [Natsuno, 2003; Paavilainen, 2001].

Third, the mobile *penetration rate* may matter. The entry timing at different mobile penetration rates may be critical for the diffusion of mobile commerce. Due to the unavoidable switching costs accumulated by incumbents and the increasingly crowded mobile market, operators have to actively push subscribers to cannibalize themselves and upgrade, as well as search more intensively for unexploited segments such as more innovative young subscribers and debuts. In 1999, the 2G mobile system in Japan was still in a growth stage and served only 40 million subscribers with a penetration rate of less than 30%. That means the mobile operators there and then still faced a large potential share before reaching the crowded competition. Meanwhile, DoCoMo targeted the young group to introduce their first time of Internet experience. Therefore, DoCoMo was able to diffuse the i-mode mobile business without confronting heavy switching costs, in contrast to its overseas licensees, located in the crowded markets with over 60% penetration. Mobile internet services are especially difficult to launch in areas where users have extensive experience with fixed-line Internet.[j] More incentives are needed to attract subscribers.

---

[j]The Northern and Western European areas experienced a higher rate of fixed-line Internet penetration than Japan before 2000. The Internet coverage in the Asia four tigers, i.e., Singapore, Hong Kong, Taiwan, and Korea, was also higher than in Japan. Japan had only 20 million Internet users, a 16.5% rate of national penetration, with only 3 million to 4 million believed to be accessing the Internet from their homes. Hence, leisure time could be further exploited for Internet access through the mobile devices, familiar to most new generation Japanese. See Hung [2000]. Also refer to "Amazing DoCoMo" and "What's It All About: An i-mode Primer" in BusinessWeek on January 17, 2000, available at http://www.businessweek.com/archives/2000/b3664010.arc.htm.

Fourth, the *coordination capability* among complements may matter. NTT, the mother of DoCoMo, has been an industry leader for many decades, along with several key Japanese electrical giants such as NEC, Hitachi, Mitsubishi, Toshiba, and Fujitsu. Industrial policies initiated by MITI, such as the 5G computer program and the SIGMA software project, have enabled cooperation among these large firms [Baba *et al.*, 1996]. This experience of collaboration has facilitated the launch of the new mobile Internet, especially in technological support and device manufacturing. Additionally, the trust between these special MITI-centric networks has also helped resolve possible wait-and-see hesitation in the face of the dispersed mobile protocols of the vertically disintegrated 2G environment. Thus, this path-dependent capability lets DoCoMo successfully coordinate the activities of several complementary equipment vendors in launching new mobile services, in contrast to the overseas licensees, who possess no such relationships. The latter depended solely on help from the *foreign* DoCoMo, rather than the local or European vendors. In sum, Japan's mobile market offers institutional advantages that have driven the transformation of mobile Internet services and ameliorated the difficulties caused by the disintegrated mobile structure.

3.3. *Unexpected deficiencies*

The first five i-mode licensees not only lacked the institutional levers but also made two important mistakes that led to a failure to stimulate the development of network effects, and to overcome the problem of switching costs.

First, i-mode vendors in Europe lacked a *differentiated product family* of mobile phones. Even though DoCoMo invited well-known Japanese device vendors such as NEC and Toshiba to support the international diffusion, they were not as familiar to western European subscribers as Nokia, Ericsson, Siemens, Alcatel, or Sagem. The lack of support from reputable European manufacturers led to not-invented-here doubts that may have delayed the diffusion of i-mode.[k] Moreover, because mobile phones are more like fashion goods than mere information appliances, licensed i-mode operators have to diversify i-mode devices, just as in the mainstream GSM/GPRS device market, where there is a diversity of mobile phones from worldwide manufacturers.[l] The fundamental problem was the Japanese standard PDC-P, which differed from the globally common GPRS standard. Many Japan manufacturers could not export their products directly to Europe. As a result, western European operators have to promote i-mode more enthusiastically if they receive support solely from DoCoMo and Japanese manufacturers. However, DoCoMo failed to implement the product proliferation strategy successfully in Western Europe or Taiwan, unlike it did in Japan.

Second, DoCoMo lacked a *more aggressive penetration pricing strategy*. As Table 3 shows, DoCoMo actually promoted i-mode services more aggressively at a lower promotion price (as deflated by average national income) than the other licensees did. That means DoCoMo subsidized subscribers with a relatively lower-priced mobile phone, providing a cheaper subscription fee, even though Japan has a higher national income than in some western European countries and Taiwan. Thus,

---

[k]Things have been going better since the end of 2003 when Nokia and Siemens decided to roll out-mode models for Germany's E-plus operator, after Samsung made a similar decision. This approach will increase German i-mode subscribership, then at only 0.14 million. In addition to Japanese manufacturers like NEC, Mitsubishi, Vitelcom, Toshiba, and Panasonic, new i-mode phones from Samsung, LG, and Sagem, the non-Japanese vendors, have been available since 2004. Refer to Reinhardt and Kunii, "Europe Is Going Mad for i-mode," BusinessWeek, July 28, 2003, available at http://www.businessweek.com/magazine/content/03_30/b3843139_mz034.htm and http://www.nttdocomo.com/presscenter/pressreleases/2003.html, /2004.html, /2005.html.

[l]The more unwanted threats from several competing technologies have crowded the few i-mode devices into the fringe of the international mobile markets. Java and MMS are supported by many well-known vendors like Nokia, Motorola, Samsung, Siemens, Sony, and Ericsson, making GSM/GPRS mobile phones more fashionable since 2000.

Table 3. Comparison of promotion programs among international i-mode operators.*

Country	Japan	Germany	The Netherlands	Taiwan	France
Cooperating operator	NTT DoCoMo	E-plus	KPN Mobile	KG Telecom	Bouygres Telecom
Gross national income per person (US$)/relative ratio	31678/1	22358/0.71	23699/0.75	12856/0.41	22307/0.70
Min. bundled price (US$)/ key i-mode mobile phone	115/A series of product family	98/(NEC n21i)	238/(NEC n21i)	144/(T535i)	238/(NEC n21i)
Fixed subscription fees (US$)	2.56	3	3	4.9	3
Variable fee for transmission (US$)	**0.017/1,000 bytes**	**0.01/1,000 bytes**	**0.002–0.025/1,000 bytes**	**0.006/1,000 bytes**	**0.01/1,000 bytes**
Mobile email fees (US$)	0.06/200 words	0.026 per mail	0.063/200 words	0.06/200 words	0.063/200 words
Content subscription fees (US$)	0.88–2.65	0.26–1.76	0.26–1.76	0.6–2.65	0.26–1.76
Relative promotional price (deflated by national income) (US$)/relative pricing ratio§	124.1/**1**	147.7/**1.19**	328.3/**2.65**	368.8/**2.97**	351.7/**2.83**

Notes: *Due to a lack of BASE-related data, this table compares the promotional price mix of only five operators.
§These relative prices are computed by deflating the total subscription costs, which were composed of the bundled price of the i-mode phone, the fixed subscription fee, and the transmission fee for 3,000 packets (128 bytes per packet) per month in the average Japanese usage level, over the gross national income per person, respectively.
Source: Adapted from http://www.nttdocomo.com, http://www.fetnet.net/, http://www.eplus.de/, http://www.kpn.com/, http://www.bouyguestelecom.fr/, http://unstats.un.org/.

i-mode licensees, in contrast to DoCoMo, did not apply this marketing strategy to effectively overcome the problem of locating in places lacking institutional levers for mobile upgrading. So, it has been slow to amplify network effects and cross the diffusion chasm.

4. Implications and Suggestions

The foregoing discussion offers deep insight into international technology transfer. Total i-mode subscribership of the first five operators within the first two years was just one-tenth of the Japanese total (Table 1). The first implication of our discussion is that the market position and the user scale truly matter, especially when a firm intends to establish a dominant design or a pioneering standard. Thus, relatively small or minor-positioned firms in the fragmented markets face more adverse institutional environments for diffusing new standards. One approach to solve the problem could be borrowed from DoCoMo's energetic expansion to 14 countries continuously across Europe and Asia. Aggregating smaller fragments into a larger scale market may be economical enough to induce the cooperation of complement providers.

The second implication is the difficulty of cross-platform innovation. In the case of i-mode's international diffusion, DoCoMo intended to extend its own mobile service protocol, crossing from the PDC-P platform to GPRS or 3G. However, the latter platform is controlled by other telecom giants such as Nokia and Lucent. Even though DoCoMo possesses special coordination capabilities and a good reputation, unless it obtains the support of other telecom giants, it will be difficult to implement the cross-platform strategy to transfer the technology to a different platform. Therefore, the reluctance or deterrence of other telecom giants, who support the open WAP protocol, may delay the migration speed of i-mode in the overseas markets where other telecom firms are dominant. In such a situation, operators will have the choice of bettering on a proprietary but successful standard or an open but lagging one. If they choose the former, they may quickly launch innovative services but be locked into a dominant provider and a proprietary standard. And if they choose the latter, they may stagnate in a wait-and-see situation until the technology matures, after which a period of intense competition will begin. Operators thus face serious problems in choosing standards.

Our study offers recommendations on business strategy and regulation policy. International expansion investment or strategic collaboration between telecom operators has become the new weapon in global competition, especially in the increasingly crowded and competitive mobile service industry [Steinbock, 2003]. UK Vodafone is a typical case.[m] However, the *proper selection of overseas target*s may increase the possibility of success for the firms to establish dominant positions in the mobile technology evolution. Global expansion targets should be able to support strategic institutional levers sufficiently, such as the large market scale and the cooperator of leading position and higher technology absorptive capacity, to foster more adopters and nurture providers of complementary applications. In other words, technological innovators and investors will come in early if prospects are bright.

The second suggestion is in carrying out a mobile strategy concerns a shift of business mindset. The mobile industry is not only an infrastructure business but also a *client-and-server*-based service. Hence, the concept of the product as a whole must be planted in the strategic operations of international mobile operators. The gateway role of mobile operators as the central servers should compel them to take responsibility for total solutions, which range from the server-end installation of equipment and aggregation of application content, to the client-end advanced mobile devices. The former will influence the utility of

---

[m]The most globalized Vodafone group included Vodafone in UK, Mannesmann in Germany, AirTouch in USA, and J-phone in Japan [Steinbock, 2003]. The Vodafone Group gained more than 67% of its revenues overseas.

mobile data transmission, while the latter will determine the network effects on subscribers. In addition, the *penetration pricing* and *product proliferation strategy* used in the latter client-end devices are necessary for creating an economically viable newly-installed base. If pioneering operators want to develop a leading platform of mobile service, the spread of client devices with higher network effects will feed back to the server in the form of a large volume of mobile transactions. A similar situation occurred with the ubiquitous PCs with compatible Wintel architecture and the rapid growth of the Internet.

For the industry policies, government regulation can either drive or hinder the diffusion of a new standard. The allowance of integration and mergers between operators, or among mobile equipment vendors and operators, may enable consolidation that will enable firms to overcome the problem of market fragmentation and take advantage of scale economies. Thus, public policy may ameliorate the difficulties of structural disintegration. Moreover, if industry policy allows a bundling strategy between operators, equipment vendors, and device retailers for developing a penetration pricing strategy, new technology innovation may diffuse easily and quickly. Additionally, telecom policy could, in contrast, restrict the operators to bundle old and less functional versions of mobile devices. This approach will equate the prices of new-bundled products with non-bundled old ones, to promote novel products to subscribers with an attractive *P/P* ratio (the functional performance over the purchase price). In addition, this policy will force the device vendors to cannibalize product lines toward the advanced versions suited for new mobile services as soon as possible. If so, the lack of diverse i-mode devices in the initial stages of international diffusion would be solved.

5. Conclusions

According to the perspectives of institutional economics, different institutions possess different dependent paths, which in turn need different self-reinforcing mechanisms for increasing returns. This study has investigated the international diffusion of i-mode from Japan to Europe by studying secondary sources such as industry archives and reports. The dominant position and the technological leadership possessed by DoCoMo in Japan were not replicated in its overseas operators. As a result, pan-western European and Taiwan operators need more effective marketing efforts from local operators and allied complementary vendors. Owing to the competitive post-2G mobile industry, the increasing degree of vertical and horizontal disintegration has rendered network effects more sluggish, since individual mobile players face higher switching costs. We unearth meaningful institutional levers that facilitate mobile transformation, and derive technology marketing implications and business strategies, as well as public telecommunications policies, that will help advance mobile technology diffusion.

In the future, monitoring the performance of i-mode global expansion, especially the entry into the unexploited but enormous Russian market and the alliance with the first operator MTS, will be an opportunity to test the validity of our arguments about institutional levers in this research. Verification of the performance of the product proliferation strategy with local and reputable vendors would be another research direction, which will become available when DoCoMo initiates equipment manufacturing collaboration with worldwide branded firms like LG, Samsung, and Motorola. In addition, different profiles of consumer preferences in mobile services may also influence the adoption of new technology beyond the institutional considerations. As Beck and Wade [2003] argued, Japanese subscribers viewed the mobile Internet as a tool of social connections rather than as a commerce utility, like subscribers in the US and Western Europe. Research on the differences in i-mode subscriber expectations among different subscriber groups may reveal more information about technology marketing and consumer behaviors.

References

Abernathy, W. J. and Utterback, J. M. (1978). Patterns of industrial innovation. *Technology Review*, **80**:40–47.

Arthur, W. B. (1988). Self-reinforcing mechanisms in economics. *The Economy as an Evolving Complex System*, Eds. P. W. Anderson, K. J. Arrow and D. Pines. Addison Wesley, MA.

Baba, Y., Takai, S. and Mizuta, Y. (1996). The user-driven evolution of the Japanese software industry: The case of customized software for mainframes. *The International Computer Software Industry: A Comparative Study of Industry Evolution and Structure*, Ed. D. C. Mowery. Oxford University Press, NY.

Beck, J. and Wade, M. (2003). *DoCoMo: Japan's Wireless Tsunami.* AMACOM, NY.

Christensen, C. M. (1997), *Innovator's Dilemma: When New Technologies Cause Great Firms to Fail.* Harvard Business School Press, MA.

Collis, D. J., Bane, P. W. and Bradley S. P. (1997). Winners and loser: Industry structure and the converging world of telecommunications, computing, and entertainment. *Computing in the Age of Digital of Convergence*, Ed. D. Yoffie. Harvard Business School Press, MA.

Funk, J. L. (2001). *The Mobile Internet: How Japan Dialed up and the West Disconnected*. ISI Publications, Pembroke, Bermuda.

Funk, J. L. (2002), *Global Competition Between and Within Standards: The Case of Mobile Phones*. Springer, Palgrave, NY.

Gawer, A. and Cusumano, M. A. (2002). *Platform Leadership: How Intel, Microsoft, and Cisco Drive Industry Innovation*. Harvard Business School Press, MA.

Hung, C.-L. (2000), Architecture competition and diffusion mechanism of system products — A case of wireless communication system. Doctoral Thesis, Institution of Business Administration of National Taiwan University, Taiwan.

Kalakota, R. and Robinson, M. (2001). *M-Business: The Race to Mobility*. McGraw-Hill.

Kotler, P., Siew, M. L., Ang, S. H. and Tan, C. T. (1996). *Marketing Management — An Asian Perspective*. Prentice-Hall.

Laffont, J.-J. and Tirole, J. (1999). *Competition in Telecommunications*. MIT Press, MA.

Lera, E. (2000). Changing relations between manufacturing and service provision in a more competitive telecom environment. *Telecommunication Policy*, **24**:413–437.

Maitland, C. F., Bauer, J. M. and Westerveld, R. (2002), The European market for mobile data: Evolving value chains and industry structures. *Telecommunication Policy*, **26**:485–504.

Matsunaga, M. (2000). モード事件. Kadokawa Shoten Publishing Co., Tokyo.

Moore, G. A. (1995). *Inside the Tornado: Marketing Strategies from Silicon Valley"s Cutting Edge*. Harper Business, NY.

Natsuno, T. (2003). *i-mode Strategy*. John Wiley & Sons.

North, D. C. (1990). *Institutions, Institutional Change, and Economic Performance*. Cambridge University Press.

Oliver, C. (1997). Sustainable competitive advantage: Combining institutional and resource-based views. *Strategic Management Journal*, **18**:697–713.

Paavilainen, J. (2001). *Mobile Business Strategies: Understanding the Technologies and Opportunities*. Wireless Press, London.

Quigley, P. (2002). Dutch courage: i-mode arrives in Europe. *Wireless Week*, **8**(1):17.

Rogers, E. M. (1995). *Diffusion of Innovations*, 4th edn. The Free Press, NY.

Rohlfs, J. H. (2001). *Bandwagon Effects in High-Technology Industries*. MIT Press, MA.

Ratliff, J. M. (2002). NTT DoCoMo and its success: Origins and implications. *California Management Review*, **44**(3):55–71.

Sabat, H. K. (2002). The evolving wireless value chain and market structure. *Telecommunication Policy*, **26**:505–535.

Scherer, F. M. (1990). *Industrial Market Structure and Economic Performance*, 3rd edn. Rand McNally & Co, Chicago.

Shapiro, C. and Varian, H. R. (1998). *Information Rules: A Strategic Guide to the Network Economy*. Harvard Business School Press, MA.

Steinbock, D. (2003). *Wireless Horizon: Strategy and Competition in the Worldwide Mobile Marketplace*. AMACOM, NY.

Teece, D. J. (1987). Capturing value from technological innovation: Integration, strategic partnering, and licensing decisions. *Research Policy*, **15**:285–305.

Vacca, J. R. (2001). *i-mode Crash Course,* 1st edn. McGraw-Hill.

Yoffie, D. B. and Cusumano, M. A. (1999). Judo strategy: The competitive dynamics of internet time. *Harvard Business Review,* **77**(1):71–81.

Chapter 6

International Perspectives: Case of Japan

Qazi Mahdia Ghyas and Fumiyo N. Kondo

1. Introduction

López-Nicolás *et al.* [2008] defined mobile information services (MIS) as advanced data services that have look and feel at internet pages and are accessible via mobile phones at 3G+ mobile communication networks. The consumers pay for their desired mobile contents or services. They may not pay additional fees for the access to excessive services. These services include mobile searches, news and sports information, music and video downloads, e-mail, instant messages, etc. Mobile information system is usually embedded in the user's life [Tamminen *et al.*, 2004]. The performance of a traditional information system (IS) is usually measured in terms of how effective it is at achieving the specific need and goals [Gefen *et al.*, 2003] or how satisfying it is to use for particular tasks [e.g., Bhattacherjee and Premkumar, 2004]. One of the goals of any technology should be to improve the quality of its users' lives [Straub and Watson, 2001]. The needs and uses of MIS may differ on the basis of age. If there are any differences present in service needs between the young and old, mobile companies need to vary their international marketing strategies and tactics.

Prior studies of mobile computing have employed satisfaction [Choi *et al.*, 2005] and intention to use [Yang, 2005] to investigate the impact of mobile computing technologies. These outcome variables measure users' feelings or attitudes at the time they use the system, rather than the impact of the system on their overall quality of life (QoL). Very few studies directly address the impact of IS on overall QoL. Rahman *et al.* [2005] presented a picture of conditions of the world with respect to such interrelated domains of QoL. Choi *et al.* [2007a] proposed 11 relationships between Individual Contribution and Overall Contribution of QoLin Korea and Japan, and found some of them had stronger influence on the Overall Contribution.

There are also very few studies on how young people's QoL has been improved owing to the mobile technology. At present, we have little sense of how much MIS affects youth's QoL, because, as said, QoL has largely been ignored in mainstream IS literature. At present, we do not even know what domains of a young user's life have been affected by QoL. Few studies have actually developed to analyze the relationship between technology and QoL. The main goal of this study is to construct a theoretical model that can reliably and validly measure the relationship between MIS and QoL. We designate "contribution to QoL" as an alternative outcome variable for MIS and examine through a consecutive empirical study in Japan, whether and how MIS contributes to young users' QoL.

Therefore, we examine the contribution of life domains to students' "QoL" after using MIS for Japan which has an expanding mobile market. We ground our research framework in the following two parts. First one is the bottom-up spillover theory (explained in Section 3.3) by Choi *et al.* [2007a] with their 11 individual life domains contribution of MIS to QoL. These life domains are: leisure life, family life, friend life, cultural life, work life, community life, consumer life, financial life, health life, safety life, and self-life. Second, our newly proposed four individual life domains contribution of MIS to QoL: home-healthcare life, informational life, educational life, and trustworthy life. These have conceptualized from previous studies [Choi *et al.*, 2007b; Jing and Andy, 2010; Nguyen *et al.*, 2010]. More details will be provided later.

The rest of the chapter is organized as follows. The next sections will describe the overview of QoL, individual life domains with new ones, bottom-up spillover theory, and contribution to QoL. These are followed by sections on methodology and statistical results. The last sections present a summary of the discussion, conclusion, implications, and direction for future research.

2. Mobile Information Services in Japan

In Japan 2014, there were 137.9 million mobile subscribers, which was 108% of total population. Among them, 108.8 million were third generation or fourth generation mobile subscribers, which was 85.3% of total population [Telecommunications Carriers Association Japan, 2014]. The smart phone market in Japan has expanded in 2010, sparked by the introduction of iPhone 4, innovative mobile services, and massive TV commercials. Japanese advertising and PR firm, Hakuhodo, surveyed internet users in Japan in June 2013 and found smartphone penetration rate is higher among younger populations. More than three-quarters of internet users aged 15–19 owned a smartphone, as did more than seven out of 10 of those between 20 and 29 years old. The Japanese people who purchased a smartphone from April to September of 2013 were 82% [Mobile marketing Data Lab, Japan, 2013].

Four factors of mobile services have identified by Ghyas *et al.* [2011]: information intensiveness (Factor 1), entertainment (Factor 2), low penetration service (Factor 3), and communication service (Factor 4), on Japan's 2008 and 2009 data. Factor 1 refers to services that require a high degree of information such as radio, Internet, learning, health, infotainment content, stock trading, shopping services, coupon and advertising information services, online storage services, reservation or booking, and location-based services. Many Japanese people look at their mobile device as a central source of information gathering. This leads to a — *Keitai* (mobile phone) culture that is more obvious in Japan than in other countries, partially because of the Japanese people's affection for technology in general. Japanese adults and teenagers rely on their mobile phones for the types of functions that a laptop or desktop computer would normally provide.

3. Theoretical Background

3.1. *Quality of life*

The term QoL has been defined differently by various scholars. McCall [1975] defines QoL as "necessary conditions for happiness", while Terhune [1973] defines it as subjective well-being. There are many terms that are used to represent well-being. Commonly used terms are QoL, standards of living, human well-being, and welfare. QoL is a measure of how happy people or how fulfilled they are in terms of their various wants and needs.

Studies of QoL have been conducted in diverse areas, including marketing (e.g., Sirgy [1996]), health (e.g., Leung *et al.* [2004]), and education (e.g., Huebner and Gilman [2002]). However, only a few studies have been conducted in the area of telecommunication services for young users. Here, we will try to measure the contribution of MIS empirically and will provide a theoretical model that explicitly links MIS and QoL for young users in Japan.

3.2. *Individual life domains*

QoL researchers have identified a number of distinct life domains that encompass the various places, things, activities, roles, and relationships in which a person typically finds himself or herself involved [Andrews and Withey, 1976]. They propose that people actually experience and store the various events of their life in distinct domains, including leisure life, family life, friend life, cultural life, work life, community life, consumer life, financial life, health life, safety life, and self-life. These 11 life domains have been examined for mobile data services (MDSs) by Choi *et al.* [2007a] concerning Japan and Korea. Our newly added life domains are educational life [Andrews and Withey, 1976; Choi *et al.*, 2007b], home-healthcare life [Jing and Andy, 2010], trustworthy life [Jing and Andy, 2010], and informational life [Nguyen *et al.*, 2010], which are discussed in the following.

- **Informational life.** Information communication technology and the Internet can offer benefits in addressing their information needs [Nguyen *et al.*, 2010]. Mobile phone with advanced technology may be a general source of information. With the availability of mobile devices, it becomes easy to gain access to the tremendous amounts of information on the Internet anywhere and anytime. So, here we will observe whether MIS may contribute to user's informational life to improve the QoL.
- **Educational life.** Word translation anytime and anywhere, reading e-books via mobile, etc., can improve the learning and teaching activities of mobile users. Choi *et al.* [2007b] did not get a significant effect between the contribution of MDSs to educational life and the contribution to overall. According to Diener *et al.* [1999], small but significant correlations between education and subjective well-being have often been found. It is to be expected that MIS may affect the young peoples' educational life to improve their overall QoLs.
- **Trustworthy life.** There are two main actors in trust: the trustor would be the consumer of MIS and the trustee, which is the trusted party. In order to provide personalized services, mobile services provider may need to locate users' position and collect users' profiles. Users may feel threatened when the technology has the capability to track users' actions and store personal information outside the users' control. Trust can help reduce the uncertainties a user faces when using mobile services. Thus, here we will observe

whether trustworthiness from mobile service provider, can improve the QoL of users.

- **Home-healthcare life.** Home-healthcare life is the activities pertaining to some self-medical care activities in home using advanced mobile technology without going to a hospital, doctor, or clinic. Users require special health vigilance, including monitoring of their physical conditions (e.g., blood pressure, etc.) and the fulfilment of some medication agenda. For example, there are some fitness applications in smartphone that are mainly designed for the young generation with the specific focus on weight loss and diet management.

We are interested here to understand how MIS contributes to QoL than in how generally satisfied people are with their lives. As the use of the mobile phone blurred the boundary between home and work [Haddon, 1998], MIS can be used crossed time and place [Palen, 2002]. Therefore, MIS has the potential to affect many life domains. Moreover, we did not know yet which domains were most relevant to MIS in different countries and different users. Therefore, our study has investigated the following life domains from Choi *et al.* [2007a], which have been shown in Table 1.

3.3. *Bottom-up spillover theory*

The bottom-up spillover theory [Andrews and Withey, 1976] is a model of the relationship between individual life domains and quality of overall life. The theory indicates that QoL in individual domains has spillover effects on overall QoL. In other words, happiness of insubordinate individual life domains can spill over to produce superordinate overall happiness [Andrews and Withey, 1976]. The premise of this model is that overall life satisfaction is functionally related to satisfaction within each of the individual life domains, which can be measured by satisfaction with specific events in each life domain [Sirgy, 2002].

Table 1. Life domains.

Life domains	Defining activities, circumstances, events
Cultural	Activities and relationships with cultural richness
Leisure	Non-working activities, spare time activities, recreation
Educational	Learning and teaching activities
Consumer	Purchase, preparation, consumption, possession, maintenance, and disposition activities of goods and services
Financial	Activities for pay and revenues
Health	Activities pertaining to mental and physical health
Safety	Activities for personal safety
Family	Activities with parents, children, and home
Friend	Activities with colleagues and friends
Self	Activities for self-representation and self-efficacy
Community	Activities with online community
Health monitoring	Activities to monitor own health care
Trustworthy	Activities of privacy and trustworthy
Informational	Activities to get information
Work	Mental and physical activities required by jobs and task

3.4. *Contribution to QoL*

The basic premise of the bottom-up spillover theory is that satisfaction levels within individual life domains affect the overall level of life satisfaction [Sirgy, 2002]. This relationship between Individual Contribution and Overall Contribution is displayed as our research model in Figure 1. The bottom layer of our model represents the contribution of MIS to QoL in those domains (henceforth, Individual Contribution). The contribution of MIS to quality of individual life domains can be measured by the

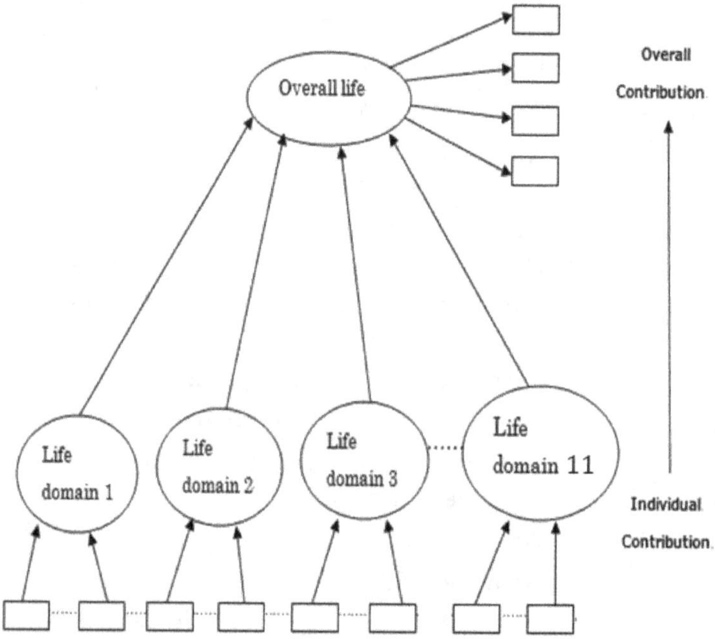

Figure 1. Relationship between individual contribution and overall contribution from Choi *et al.* [2007a].

perceived contribution of specific use-experience of MIS services in that domain. The top layer in our model represents the specific contribution of MIS to overall QoL (henceforth, Overall Contribution). Like Individual Contribution, Overall Contribution is conceptualized as the perceived contribution of MIS services to overall QoL.

Past studies suggest that IT does have influence on users' QoL. For example, Heijden's [2004] study of adoption behavior for hedonic systems implies that IT may have effects on leisure or culture life. Bhattacherjee's [2001] account of users' continuance behavior in online banking systems implies that such systems affect QoL in the domain of financial life. Choi *et al.* [2007a] measured the contribution of MDS technology to the quality of users' lives. They have examined 11 individual life domains for Korean and Japanese respondents. Whereas,

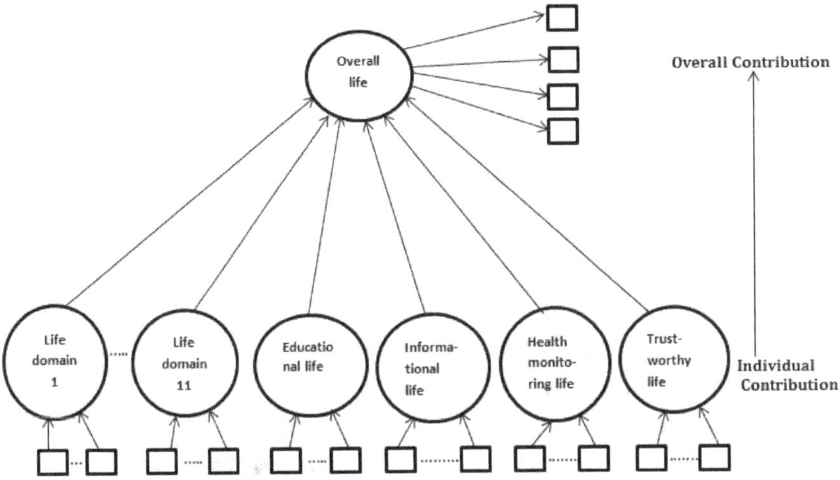

Figure 2. Relationship between individual contribution and overall contribution with our proposed life domains.
Source: Life domains from Choi *et al.* [2007a, 2007b, pp. 1–11].

we will try to examine 15 life domains with our proposed four life domains.

Our measures concern spillover of contribution to QoL by MIS. Therefore, contributions of MIS to QoL within individual domains should affect the contribution of MIS to overall QoL. Our proposed four new life domains are displayed as our research model in Figure 2. We expect that improvement of QoL in each life domain (original and additional) after using MIS will positively influence overall QoL for young Japanese mobile users.

4. Methodology

4.1. *Data*

A survey was conducted in Japan for 245 mobile phone users at a university in Ibaraki. An online questionnaire was distributed. The Japanese survey was conducted in Ibaraki in February 2013.

We used responses from users who were 20–39 years for this research. Major respondents were young adults, as it has been found that the mobile phone has become a key information technology for youth [de Chenecey, 2002]. Our participants are the Japanese young users of MIS via mobile phone with internet convergence. For having Internet in mobile phone, most of the users have smartphone. So, the sample in our research contains mostly smart phone users. We eliminated some responses which had missing values. Therefore, after the elimination, in Table 2, there were 189 respondents to do the analysis for Japan, 2013.

Questions measuring Individual Contribution in the 15 life domains were based on the MIS use experiences. Our initial survey items for 11 original life domains were adapted from Choi *et al.* [2007a]. The items of our additional four life domains were adopted from Andrews and Withey [1976], Jing and Andy [2010], and Nguyen *et al.* [2010]. We chose to use the formative indicators elicited in order to reflect concrete use experiences with MIS because our theoretical foundation, the bottom-up spillover theory, required highly specific events and experiences at the bottom of the satisfaction hierarchy model. Subjects were asked to rate how much they agree that each specific experience of using MIS contributes to the QoL in the corresponding individual life domains. We constructed questions measuring Overall Contribution of MIS to QoL as reflective indicators based on the Satisfaction with Life Scale (SWLS), which is one of the most widely used scales in QoL

Table 2. Demographics of respondents.

Category	Values	Japan 2013	In %
Gender	Male	137	72
	Female	52	28
Age	19 years old or younger	32	17
	20–29 years old	151	80
	30–39 years old	6	3

research, with high internal consistency and test–retest reliability [Pavot and Diener, 1993]. We used the following sets of 7-point Likert scales: strongly disagree/strongly agree for Individual and Overall Contributions. The survey questions are shown in Appendix.

4.2. *Analytical method*

The validity and reliability of the measures were tested using Partial Least Squares (PLS), a method well suited to handle both formative indicators (for Individual Contribution) and reflective indicators (for Overall Contribution) [Chin, 1998]. The research models were analyzed for Japan 2013 via PLS by using the statistical software SmartPLS Version 2.0 M3. PLS also has been used to construct a theory of MDS contribution to QoL by Choi *et al.* [2007a] concerning Japanese and Korean mobile users.

4.3. *Validity and reliability of the measures*

There are two types of measurement scale in structural equation modeling, i.e., formative or reflective. In the research by Choi *et al.* [2007a, 2007b], both reflective and formative measurements were present in the same model. PLS is appropriate for handling both formative indicators and reflective indicators [Chin, 1998].

Because the criteria for assessing reflective and formative constructs are different, we assess the two types of constructs separately. In PLS, loadings represent the influence of individual scale items on reflective constructs [Bollen and Lennox, 1991]. For reflective measurement scale, their indicator reliability, internal consistency reliability, or discriminant validity should be fully examined [Hair *et al.*, 2013]. Therefore, in order to assess the reliability and validity of the reflective constructs, we checked the factor loadings as confirmatory factor analysis, composite reliability (CR), Cronbach α coefficient, and average variance extracted (AVE) for Overall Contribution in Table 3. The results indicate that all the measuring items secure convergent

Table 3. Reliability and validity of overall contribution.

Reflective indicator	Items	CR ≥0.70	AVE ≥ 0.50	Cronbach $\alpha \geq 0.70$	Loading ≥ 0.50	*t*-value at $p < 0.01$
Overall contribution	OVR 1	0.93	0.78	0.90	0.91	90.18
	OVR 2				0.93	108.74
	OVR 3				0.93	108.73
	OVR 4				0.75	25.27

Table 4. Collinearity statistics of formative constructs.

	Collinearity statistics	
Independent variables	Tolerance > 0.20	VIF < 5
Community life	0.552	1.810
Consumer life	0.449	2.228
Cultural life	0.524	1.909
Educational life	0.453	2.207
Family life	0.516	1.939
Financial life	0.546	1.832
Friend life	0.470	2.127
Health life	0.475	2.103
Home-healthcare life	0.480	2.083
Informational life	0.363	2.757
Leisure life	0.385	2.597
Safety life	0.405	2.467
Self-life	0.537	1.864
Trustworthy life	0.389	2.567
Work life	0.455	2.196

Note: Dependent variable: Overall contribution.

validity at the indicator level, and construct level, and the internal consistency.

Regarding the formative construct, In PLS, weights represent the comparable influence on formative constructs [Bollen and Lennox, 1991]. Thus, we examine the formative item's weights, multicollinearity, and discriminant validity of the formative construct. In a formative measurement model, the problem of indicator collinearity may occur if the indicators are highly correlated to each other. We have generated VIF and tolerance values for collinearity checking. The collinearity statistics, as shown in Table 4, indicate that multicollinearity is not severe [Hair *et al.*, 2013]. For confirmatory factor analysis in Table 5, the measuring items for one formative construct have the same factor loadings. All the individual measuring items

Table 5. Reliability and validity of individual contribution.

Formative indicators	Items	CR ≥ 0.70	Weight	*t*-value	Factor loadings ≥ 0.50	*t*-value
Health life	HEL 1	0.87	0.17	1.48	0.68	7.75
	HEL 2		0.64	5.14	0.94	23.09
	HEL 3		0.36	2.53	0.81	11.06
Home-healthcare life	HML 1	0.92	0.04	0.23	0.63	5.91
	HML 2		0.98	9.57	0.99	79.28
Work life	WOR 1	0.86	0.56	5.72	0.87	16.77
	WOR 2		0.12	1.47	0.67	11.2
	WOR 3		0.51	4.69	0.84	13.96
Safety life	SAF 1	0.83	0.67	6.68	0.94	27.34
	SAF 2		0.26	2.61	0.79	13.3
	SAF 3		0.28	3.37	0.58	7.65
Financial life	FIN 1	0.91	0.28	1.98	0.79	12.27
	FIN 2		0.19	1.18	0.8	11.17
	FIN 3		0.44	3.43	0.8	11.6
	FIN 4		0.34	2.18	0.82	12.92
Consumer life	CSR 1	0.87	0.14	1.23	0.72	8.76
	CSR 2		0.32	2.85	0.8	12.38

	CSR 3		0.63	5.37	0.94	26.15
	CSR 4		0.1	1.08	0.56	6.97
Family life	FAM 1	0.87	0.34	3.06	0.74	10.63
	FAM 2		0.31	2.41	0.81	13.12
	FAM 3		0.46	3.64	0.82	14.61
	FAM 4		0.15	0.97	0.76	10.44
Cultural life	CUL 1	0.92	0.49	3.79	0.83	14.33
	CUL 2		0.35	2.58	0.87	20.02
	CUL 3		0.06	0.55	0.9	23.88
	CUL 4		0.08	0.47	0.75	13.78
	CUL 4		0.19	1.9	0.73	12.56
Friend life	FRI 1	0.9	0.19	1.7	0.76	12.93
	FRI 2		0.69	6.05	0.97	42.36
	FRI 3		0.24	2.12	0.79	12.6
Community life	COML 1	0.94	0.57	4.04	0.95	29.68
	COML 2		0.5	3.53	0.93	24.54
Self-life	SEL 1	0.84	0.95	16.25	0.99	66.48
	SEL 2		0.12	1.02	0.48	4.19

(Continued)

Table 5. (*Continued*)

Formative indicators	Items	CR ≥ 0.70	Weight	*t*-value	Factor loadings ≥ 0.50	*t*-value
Informational life	IL 1	0.86	0.52	5.28	0.87	19.41
	IL 2		0.57	5.49	0.59	9.23
	IL 3		0.06	0.78	0.9	21.41
Trustworthy life	TWL 1	0.86	0.81	8.75	0.98	45.16
	TWL 2		0.08	0.88	0.72	9.16
	TWL 3		0.23	1.91	0.61	7.95
Educational life	EL 1	0.89	0.5	5.85	0.89	23.54
	EL 2		0.38	3.62	0.85	21.2
	EL 3		0.29	2.56	0.8	13.65
Leisure life	LSRL 1	0.91	0.39	3.7	0.86	18.93
	LSRL 2		0.45	4.57	0.87	22.26
	LSRL 3		0.31	3.17	0.88	26.83

show a value of 0.5 or above at the significance level of 1% in our model. Thus, all the measuring items used in this study are valid to measure equivalent variables in an individual manner. We used the standard Bootstrapping procedure of 5,000 samples recommended by Hair *et al.* [2013] to test convergent validity. The weights and *t*-statistics of all formative indicators in Table 5 are found to bear significant relationships within the corresponding life domains. Moreover, the weights for each of the question items are positive, which indicates that measures of Individual Contribution also were found to have the appropriate level of convergent validity.

We tested nomological validity of the model with and without our additional four life domains. We tested whether domain-specific contribution of MIS to QoL had a significant relationship with Overall Contribution of MIS to QoL. The coefficient of determination, R^2, was 0.58 for Model 2, indicating acceptable fit in Table 6. The 15 individual life domains moderately explain 58% of the variance in Overall Contribution.

Meanwhile, predictive relevance is another aspect that can be explored for the inner model. The obtained Stone-Geisser's ($\mathbf{Q}^2$) value (i.e., cross-validated redundancy measures) for Overall Contribution was 0.42 for Model 2 in Table 6, which indicates large predictive relevance. Regarding the overall quality of the PLS structural model, we computed the value of the goodness of fit (GoF) [Tenenhaus *et al.*, 2005], which is shown in Table 6. The GoF, $\mathbf{Q}^2$, and R^2 values for Model 2 are larger than

Table 6. Model selection criteria for each model.

Model selection criteria	R^2 of endogenous construct	$\mathbf{Q}^2$	GoF
Model 1 from Choi *et al.* [2007a] (Figure 1)	0.54	0.40	0.65
Model 2 with our additional four life domains (Figure 2)	0.58	0.42	0.67

those for Model 1, resulting in our proposed model comparatively a better model. Therefore, our newly added four life domains in Model 2 are useful to forecast overall life, and Overall Contribution and Individual Contribution were found to have appropriate levels of reliability and validity for Japan.

5. Contribution of MIS to QoL for Japan

According to the theoretical model in Figure 2, the Individual Contribution of MIS will influence the Overall Contribution of MIS. In other words, if people feel that MIS improves their QoL in individual domains, they will also believe that MIS improves their overall QoL. However, we do not know which of the 15 life domains are more closely related to the overall QoL. In order to explore these issues, we conducted a PLS analysis using the Bootstrap method of 5,000 samples that is recommended by Hair *et al.* [2013] and estimated the path coefficients between Individual Contribution and Overall Contribution in Table 7.

The results from Japan indicate that contributions to six individual life domains had significant impacts on the Overall Contribution of MIS. In other words, if MIS can contribute to the QoL in the six life domains, it will have a high possibility of improving the overall QoL of young Japanese respondents. Contribution to the informational life has the highest influence on the Overall Contribution ($\beta = 0.19$, $t = 2.97$, $p < 0.001$), followed by culture life ($\beta = 0.149$, $t = 4.273$, $p < 0.001$), friend life ($\beta = 0.17$, $t = 2.70$, $p < 0.001$), leisure life ($\beta = 0.15$, $t = 2.84$, $p < 0.001$), educational life ($\beta = 0.11$, $t = 2.34$, $p < 0.05$), and work life ($\beta = 0.10$, $t = 2.02$, $p < 0.05$). As the *t*-statistics, all these path coefficients are significant at the level of 5% or less.

Our result for Japan has some similarities with those from Choi *et al.* [2007a, 2007b] for Korea and Japan. Choi *et al.* [2007a, 2007b] measured the contribution of MDS technology to the quality of users' lives. From our Japanese results, contribution to the cultural and leisure lives has the influence on the

Table 7. Comparative analysis between Japan 2013 and Japan–Korea from Choi *et al*. [2007a, 2007b] in terms of path coefficients.

Hypotheses	Japan 2013	*t*-value	Japan [Choi *et al.*, 2007a]	*t*-value	Korea [Choi *et al.*, 2007a]	*t*-value
Cultural life --> Overall life	0.18	4.08***	0.12	3.85***	0.02	0.27
Leisure life --> Overall life	0.15	2.84***	0.11	3.39***	0.06	1.63
Community life --> Overall life	0.08	1.78	0.08	2.13	0.15	4.27***
Consumer life --> Overall life	0.07	1.32	0.09	2.05	0.05	1.52
Family life --> Overall life	–0.08	1.66	0.02	0.94	0.14	3.20***
Health life --> Overall life	0.10	1.68	–0.00	0.31	0.00	0.03
Safety life --> Overall life	–0.01	0.22	0.02	0.55	0.02	0.67
Self-life --> Overall life	–0.08	1.65	0.37	10.84***	0.33	9.34***
Financial life --> Overall life	0.00	0.09	0.23	6.34***	0.14	3.87***

(Continued)

Table 7. (*Continued*)

Hypotheses	Japan 2013	*t*-value	Japan [Choi *et al.*, 2007a]	*t*-value	Korea [Choi *et al.*, 2007a]	*t*-value
Friend life --> Overall life	0.17	2.70***	0.03	0.78	0.06	1.57
Work life --> Overall life	0.10	2.02**	–0.06	2.52***	0.04	1.21
Home-healthcare life --> Overall life	0.09	1.64	—	—	—	—
Educational life --> Overall life	0.11	2.34**	—	—	—	—
Informational life --> Overall life	0.19	2.97***	—	—	—	—
Trustworthy life --> Overall life	–0.02	0.45	—	—	—	—
R^2 value	0.58		0.63		0.51	

Notes: ***p <0.001, **p <0.05.

Overall Contribution. These results are the same as those were significant in Choi *et al.* [2007a, 2007b] for Japan also. As the country is the same, i.e., Japan, we got similar results probably due to the same culture and economy. We also have some similarities in case of non-significant results. Consumer life, health life, and safety life were not the determinants of Overall Contribution in Choi *et al.* [2007] for both Japan and Korea, and we also found the same for Japan in 2012.

Our results for Japan have some differences compared to those from Choi *et al.* [2007a] for Korea and Japan. They got significant effects of financial life and self-life on the overall contribution for Japan. But, we get insignificant results on those. This may be possibly due to the characteristics of sample such as gender ratio. Another difference is on friend life, which is significant in our result, but insignificant for Japan as well as Korea in those of Choi *et al.* [2007a]. All of our respondents are students, so, friend life is a focus for them. The contribution to work life in Choi *et al.* [2007a] for Japan was found to be negatively related to the Overall Contribution but, in our result, it is opposite, positively related. Our respondents in Japan were in their twenties, so it is reasonable that work life as part-timers may be important for them. Those differences indicate the sample difference on gender, age, or job status and timing.

Out of our additional four life domains, informational and educational life domains become significant for Japan in our results. Since the respondents are all students, MIS has a strong impact on their educational lives due to searching dictionary, online translation, online education, etc. MIS also has a strong impact on their informational life due to searching news, weather, maps, personal information, etc. Our data was from 2013 at the later adoption stage of 3G/4G services, and mobile device had become a key device for information gathering after the introduction of 3G/4G services. It helps user to get information via mobile anytime and anywhere.

6. Discussion

This study has two layers (bottom layer: Individual Contribution; top layer: Overall Contribution) to measure the contribution of MIS to users' QoL. Results from the two consecutive studies clearly indicate that the two variables are valid and reliable in the bottom-up spillover theory. As users perceived a stronger contribution of MIS in specific life domains, they perceived a stronger contribution of MIS to their overall QoL. The Individual Contribution in cultural, leisure, informational, friend, educational, and work domains had the greatest impact on the Overall Contribution among Japanese young users.

Our demographic values for Japan have some differences from Choi *et al.* [2007a] concerning Korea and Japan. They had 89% Korean respondents of age group below 30 years. On the other hand, they had 17% Japanese respondents' of age below 30 years. We have 97% Japanese respondents who are below 30 years old. The age of our sample has similarities with their sample for Korea, but is different from their Japanese sample. In their study, 55–56% respondents were male for Korea and Japan. However, 72% of respondents in our study are male, so that our sample is skewed in gender. Demographic information about the effective participants is presented in Table 8. They conducted a large-scale online survey in 2005–2006 before the introduction of 3G services with diverse mobile user groups who had different technological, educational, and economic profiles. On the other hand, our online survey was conducted in 2013 among students in a university when MDSs have adopted much and also have almost the same technological, educational, and economic profiles. The difference of gender, age, time, and user group may affect our results compared to them.

Cultural and leisure life domains are significant as those in Choi *et al.* [2007a] for Japan. They explained the strong effects of leisure and cultural lives for the Japanese people by the higher national wealth (GNP and GDP) of Japan. People with greater monetary resources are more interested in their leisure

Table 8. The comparison of sample differences with Choi *et al*. [2007a].

Category	Values	Japan 2013	In %	Korea [Choi *et al*., 2007a]	In %	Japan [Choi *et al*., 2007a]	In %
Gender	Male	137	72	507	56	569	55
	Female	52	28	397	44	469	45
Age	19 years old or younger	32	17	177	20	5	1
	20–29 years old	151	80	616	69	174	16
	30–39 years old	6	3	96	11	462	45
	40 years old or older	N/A	N/A	18	2	397	38

Note: The age expression in Choi *et al*. [2007a] has been corrected as follows: From "Under 19 years" to "19 years old or younger" and from "Over 40 years" to "40 years old or older."

and cultural lives. Although our dataset was for 2013, we got a similar result because of the stable economy of the same country, Japan.

Contribution of MIS to QoL in the informational life, our newly added life domain, has a significant positive impact on Overall Contribution of MIS. According to the proposed theory by Wilson [2006], information-seeking behavior is triggered from physiological, socio-psychological, and cognitive needs as perceived by the information user. Wilson [2006] especially stressed the importance of studying users' information needs and that such needs are embedded in the user's life and world, the user him/herself, his/her role (e.g., job tasks, care needs), and the environment. The information users in Japan may seek information from different sources. Mobile device is one of the greatest devices among choices to get information and Japanese young people can improve their life by getting information anytime and anywhere through mobile devices.

Contribution of MIS to QoL in the friend life has a significant positive impact on Overall Contribution of MIS. Relationship with friends should be considered in any measure of QOL [Rahman *et al.*, 2005]. Satisfaction with friend life is an important element of an individual's well-being, in fact, even when they are young adults. It is quite reasonable that, in most cases, an individual young mobile user with strong friend ties will be happier by sending photos, contacting with friends, congratulating friends on their birthday, etc.

Contribution of MIS to QoL in the educational domain has a significant effect on Overall Contribution of MIS to QoL for Japan. But Choi *et al.* [2007b] did not get significant effects between the contribution of MDS to educational life and the contribution to overall for Korean young users. They showed the reasons were poor usability of mobile devices, which inhibits active participation, and their limited bandwidth, which precludes effective use of multimedia broadcasting on online education. Our dataset was for Japan, 2013, when around 82% of cell phones users were smart phone users. So, mobile devices become effective media for online education of young students. Hence, our result for educational life domain may become significant.

Contribution of MIS to QoL in work life has a significant positive impact on Overall Contribution of MIS in our result. Rahman *et al.* [2005] identified the work life domains depending on what were considered to be the major element of well-being. Estimates of unemployment rate and combined first-, second-, and third-level school gross enrolment ratios are used to capture the "extent of work and productive activity" that exists in a country. Young citizens of Japan are productively engaged either in work employment or in the process of learning in school.

7. Implications

The study has several implications. First, it was noted that six significant individual life domains, specifically relevant to mobile computing environments, contribute in multiple ways to various facets of a young user's life. We recommend focusing on these life domains for young mobile users. Second, it verifies the reliability and the convergent, divergent, and nomological validity of the original and our newly proposed life domains, and through an empirical study. These life domains can be used in future studies to measure MIS contribution to QoL.

This study firstly provides a framework to understand the contribution of MIS to youth's QoLs from the perspective of its actual user. The life domains proposed and verified in our study can be used in future studies, to evaluate the potential of MIS to contribute to young user's QoL. If a firm intends to specialize in a specific life domain for young users, our findings on contribution of MIS to QoL can indicate what services will be used, with what goals, in what contexts, and can thus suggest which services will most improve QoL for their users. The results on the relationship between domain-specific contribution of MIS to QoL and Overall Contribution of MIS to QoL will also help companies allocate resources to the life domains that bear most strongly on overall QoL.

8. Limitations

Our study has some limitations. All of the respondents are young students who have good education and knowledge in technology from a national university in Japan. Also, they are convenience samples. Thus, the results cannot be generalized to the whole population in Japan. Moreover, our respondents in Japan were clustered on males in their twenties which might bias the survey results. Second, this study was conducted in Japan with Japanese MIS users and it is not clear whether the results can be applied to other countries. We recommend an increased focus on the importance of age, gender, culture, and economic characteristics in future research.

9. Conclusion

This paper operationalized 15 life domains with our newly added four life domains to measure the contribution of MIS to QOL across Japanese young students. We successfully applied bottom-up spillover theory for the use of MIS by Japanese young users, and found the importance of information and educational life domains out of our proposed four life domains. By using a model to measure the contributions of MIS to QoL with the original and our additional life domains, the research of Choi *et al.* [2007a], and this research gave stable results after considering the time difference. Out of these considered life domains of the QOL, we showed that the perceived contribution of each use experience for the following six life domains were found to be the key domains for Japanese young MIS users: informational life, cultural life, friend life, leisure life, educational life, and work life.

Our research also identified the importance of appropriateness of sample characteristics such as gender, age, and culture/economy. A technology used in as many contexts as MIS can enhance its users' lives in numerous ways. To assess those contributions, however, we need to know how they focused on specific tasks and functions for different user segments. Our study takes a step in that direction by offering a practical basis for measuring the contributions of MIS to QoL for young MIS users.

Acknowledgment

This work was supported by Grants-in-Aid for Scientific Research, No. 25245054, from the Ministry of Education of the Japanese government.

References

Andrews, F. M. and Withey, S. B. (1976). *Social Indicators of Well-Being: America's Perception of Life*. Plenum Press, NY.

Bhattacherjee, A. (2001). Understanding information systems continuance: An expectation-confirmation model. *MIS Quarterly*, **3**(25): 351–370.

Bhattacherjee, A. and Premkumar, G. (2004). Understanding changes in belief and attitude toward information technology usage: A theoretical model and longitudinal test. *MIS Quarterly*, **28**(2): 229–254.

Bollen, K. A. and Lennox, R. (1991). Conventional wisdom on measurement: A structural equation perspective. *Psychological Bulletin*, **2**(110):305–314.

Chin, W. W. (1998). The partial least squares approach to structural equation modeling. *Modern Methods for Business Research*, Ed. G. A. Marcoulides. Psychology Press.

Choi, M. S., Choi, H., Kim, Y. C. and Kim, J. (2005). An empirical study on post-adoption behavior of mobile data service with a focus on comparison between services. *Korean Marketing Review*, **1**(20):1–27.

Choi, H., Im, S. K., Lee, M. and Kim, J. (2007b). Contribution to QoL: A new outcome variable for mobile data service. *Journal of the Association of Information System*, **8**(12):598–618.

Choi, H., Kim, J., Lee, M., Gaglis, G., Naruse, K., Constantious, I., Fife, E. and Mayhew E. G. (2007a). Contribution to QoL: Cross-national validation of new metrics for mobile data service technology in Korea and Japan. *International Journal of Innovation and Technology Management*, **4**(2):191–208.

de Chenecey, S. P. (2002). Putting brands in their place. *Young Consumers*, **4**(2):47–50.

Diener, E. D., Suh, E. M., Lucas, E. E. and Smith, H. L. (1999). Subjective well-being: Three decades of progress. *Psychological Bulletin*, **125**(2):276–302.

Gefen, D., Karahanna, E. and Straub, D. W. (2003). Trust and TAM in online shopping: An integrated model. *MIS Quarterly*, **27**:51–90.

Ghyas, Q. M., Kondo, N. F. and Kawamoto, T. (2011). Communication needs of Japan and the United States: A comparative analysis of the use of mobile information services. In *Proceedings of the First International Conference on Mobile Services, Resources, and Users*, Barcelona, Spain.

Haddon, L. (1998). Il Controllodella Comunicazione. Imposizione di Limitiall'usodelTelefono, *Telecomunicando in Europa*, Ed. L. Fortunati. Franco Angeli, Milano.

Hair, J. F., Hult, G. T. M., Ringle, C. M. and Sarstedt, M. (2013). *A Primer on Partial Least Squares Structural Equation Modeling (PLS-SEM)*. Sage Publications, Thousand Oaks.

Heijden, V. D. (2004). User acceptance of hedonic information systems. *MIS Quarterly*, **28**:4.

Huebner, E. S. and Gilman, R. (2002). An introduction to the multidimensional students' life satisfaction scale. *Social Indicators Research*, **60**:115–122.

Jing, G. and Andy, K. (2010). Mobile application development for senior citizens. In *PACIS 2010 Proceedings*, pp. 214–223. Available at http://aisel.aisnet.org/pacis2010/65.

Leung, K. L., Kwok, S. P., Lam, S. C., Lee, J. F., Yiu, R. Y., Ng, S. S., Lai, P. B. and Lau, W. Y. (2004). Laparoscopic resection of recto sigmoid carcinoma: Prospective randomized trial. *Lancet*, **363**:2224–2229.

López-Nicolás, C. L., Castillo, M. F. and Bouwman, H. (2008). An assessment of advanced mobile services acceptance: Contributions from TAM and diffusion theory models. *Information & Management*, **45**(6):359–364.

McCall, S. (1975). QoL. *Social Indicators Research*, **16**:299–349.

Nguyen, L., Evans, S., Wilde, W. D. and Shanks, G. (2010). Supporting information needs of senior citizens in community aged care. In *ACIS 2010 Proceedings — 21st Australasian Conference on Information Systems*, p. 7.

Palen, L. (2002). Mobile telephony in a connected life. *Communications of the ACM*, **45**(3):78–82.

Pavot, W. and Diener, E. (1993). Review of the satisfaction with life scale. *Psychological Assessment*, **2**(5):164–171.

Rahman, T., Ron, C. M. and Wandschneider, P. (2005). Measuring the QoL across countries: A sensitivity analysis of well-being indices. World Institute for Development Economics Research, UNU-WIDERWIDER Research Paper, Volume: 2005/06.

Sirgy, M. J. (1996). Strategic marketing planning guided by the quality-of-life (QoL) concept. *Journal of Business Ethics*, **15**:241–259.

Sirgy, M. J. (2002). *The Psychology of QoL*. Kluwer Academic Publishers, Dordrecht.

Straub, D. W. and Watson, R. T. (2001). Research commentary: Transformational issues in researching IS and net-enabled organizations. *Information Systems Research*, **4**(12):337–345.

Tamminen, S., Oulasvirta, A., Toiskallio, K. and Kankainen, A. (2004). Understanding mobile contexts. *Personal and Ubiquitous Computing*, **8**:135–143.

Telecommunications Carriers Association Japan (2014). Available at http://www.tca.or.jp/english/database/2010/02/index.html.

Tenenhaus, M., Vinzi, V. E., Chatelin, Y. M. and Lauro, C. (2005). PLS path modeling. *Computational Statistics and Data Analysis*, **48**(1):159–205.

Terhune, K. W. (1973). Probing policy relevant questions on the QoL. *The Quality of Life Concept*. Environmental Protection Agency, Washington, DC.

Wilson, T. D. (2006). On user studies and information needs. *Journal of Documentation*, **62**(6):658–670.

Yang, K. C. C. (2005). Exploring factors affecting the adoption of mobile commerce in Singapore. *Telematics and Informatics*, **3**(22): 257–277.

Appendix. Survey Questionnaire to Know How do MIS Improve QoL

A.1. Overall, how do you feel about MIS and the overall QoL?

Overall Contribution	Items
	Using the MIS helps me make my life close to ideal; Using the MDS improves the general conditions of my life; Using the MDS helps make my life more satisfying; Using the MDS helps me achieve important things in my life.

A.2. Original individual life domains from Choi *et al*. [2007a].

Original life domains	Items
Cultural	Purchasing movie or concert tickets over the MIS whenever I want improves my cultural life; Making reservations for movies and concerts through the MIS while I am out improves my cultural life; Getting discount vouchers through the MIS before going to a restaurant or cinema improves my cultural life; Getting movie information through the MIS on the street improves my cultural life; When it is too bothersome to go to the cinema, reserving tickets through the MIS improves my cultural life.
Leisure	Using the MIS to lift my spirits when I am gloomy improves my leisure life; Using the MIS to spend my spare time while I am out helps my leisure life; MIS help me to stay close with my personal interests, preferences, and hobbies which improves leisure life.
Consumer	Frequently checking prices through the MIS while I am shopping improves my life as a consumer; Buying goods through the MIS instead of

(*Continued*)

A.2. (*Continued*)

Original life domains	Items
	actually going out for shopping improves my life as a consumer; Searching for information on goods I want to buy through the MIS improves my life as a consumer; Exchanging goods over the MIS improves my life as a consumer.
Financial	Using the MIS to send money electronically to another person when I am away improves my financial life; Using the MIS to check my bank account improves my financial life; Using the MIS to buy and sell stocks/options instantly improves my financial life; Making reservations for trains with low commissions on the MIS improves my financial life.
Health	Recording health data daily through the MIS improves my health life; Using the MIS to take the medical advice when I cannot use a PC improves my health life; Instead of going to the clinic or hospital, making an appointment by the MIS will improve my health life.
Safety	Using the MIS to search my location when I get lost at a strange place improves my personal safety; Searching for a safe route to my destination through the MIS late at night improves my personal safety; Being in contact with other people through the MIS when I am alone improves my personal safety.
Family	When I am concerned about my family members, searching for their current location using the MIS improves my family life; Getting in touch with my family members through the MIS to relieve their worries improves my family life; Sending photos to my family over the MIS help my family life; To be in constant contact with my family when I am away improves my family life.

(*Continued*)

A.2. (*Continued*)

Original life domains	Items
Friend	Frequently sending photos to friends over the MIS improves my friendships; Frequently contacting friends through the MIS improves my friendships; Congratulating friends on their birthdays over the MIS improves my friendships.
Self	Using the MIS to share my own photos with others improves my self-expression; Immediately upgrading to the latest services over the MIS increases my self-esteem.
Community	Informing members of online community about meeting dates over the MIS improves my community life; Checking bulletin boards and e-mail from the online community over the MIS improves my community life.
Work	Searching for recruiting information through the MIS helps my work life; Discussing job-related issues with co-workers through the MIS helps my work life; Working through the MIS when I cannot use a PC helps my work life.

A.3. Proposed individual life domains.

Proposed life domains	Items
Trustworthy	Having trust on the internet services by MIS improves my trustworthy life; Having trust on paying for the online shopping via MIS will improve my trustworthy life; Having trust on MIS while I am reading, entertaining, and getting information via Internet will improve my trustworthy life.

(*Continued*)

A.2. (*Continued*)

Original life domains	Items
Information	MIS helps me to get information of news, weather, maps, etc. when I need, which improves my information life; I can search personal information through MIS when I need, which improves my personal life; At anytime, anywhere, to be searching information at anytime, anywhere by the MIS is fully useful in order to improve my informational life.
Home-healthcare	Using MIS to monitor blood pressure, glucose level, weight, food calorie, etc. by own self when I am home, improves my health life; Using MIS in home for diet control and medication agenda, etc., improves my health care.
Educational	Translate words through mobile dictionary anywhere, anytime improves my educational life; Reading newspapers, e-books through MIS improves my educational life; To see the education program by the MIS when I cannot use the PC will improve my educational life.

Chapter 7

International Perspectives: Cases of Germany and South Africa

Miriam Martens, Oliver Roll, and Roger Elliott

1. Introduction

Mobile payment (m-payment), defined as a payment procedure in which the payer uses mobile electronic techniques and devices to initiate, authorize, or make a payment [BITKOM, 2014], has been studied for more than a decade. Even so, the technology has not been extensively adopted yet.

Several theoretical models have been developed to explain the adoption of information technologies such as m-payment. One of the most frequently used models is the technology acceptance model (TAM) [Davis, 1989]. In mobile services acceptance research, TAM has been extended by various factors to explain adoption, e.g., perceived risk and compatibility [Chen, 2008] or perceived credibility, perceived self-efficacy, and perceived financial cost [Luarn and Lin, 2005]. Gao *et al.* [2014] expanded TAM by context, trust, and personal initiatives and characteristics to propose a new research model, named the mobile services acceptance model. Personal initiatives and characteristics had the strongest influence on the intention to use a mobile service, indicating that it was not just the system-specific characteristics, but the consumers' personality traits too, which play a role in adoption decisions. Besides, Agarwal and Prasad [1999] empirically tested and proved that individual differences, such as a person's role with respect to technology and their level of education, are significantly related to the variables of the TAM construct. It is, therefore, of major interest for practitioners and researchers to further investigate the relationships between individual- and system-specific dimensions and the acceptance of information systems.

The Technology Readiness Index 2.0 (TRI 2.0) measures peoples' propensity to embrace and use innovative technologies and comprises four individual-specific dimensions [Parasuraman and Colby, 2014]. The previous version, TRI 1.0 [Parasuraman, 2000], has been combined earlier with TAM, also in the context of m-payment [Guhr *et al.*, 2013]. Till date, however, it is not known how each single variable of the TRI 2.0 is related to the acceptance of m-payment and whether the factor "country" moderates the structural relationships. In addition, Guhr *et al.* [2013] recognized the need to consider newly industrialized countries while investigating how the adoption of mobile payments is influenced by constructs and cultures. The relevance of cross-country and cultural differences in the adoption of innovations has also been recognized and empirically verified by Singh [2006] and Erumban and de Jong [2006].

Hence, the overarching goal of this research project is to examine how each dimension of the TRI 2.0 affects the central dimensions of TAM for m-payments in Germany and South Africa. This comparison is particularly interesting because of the apparent differences in infrastructure and consumer readiness. Germany is characterized by low consumer readiness, but offers a solid infrastructure for m-payments. South Africa's consumer readiness is above average, but the country lags behind in terms of infrastructure [MasterCard, 2012]. However, in contradiction to these findings, Berndt *et al.* [2010] obtained a lower TRI 1.0 score for South Africa in comparison with the original study in the USA [Parasuraman, 2000], which could imply that consumers in developing countries are less ready to use innovative banking technologies.

Against this background, an empirical study was conducted at one university each in Germany and South Africa. The results of this study will help researchers to assess the applicability and invariance of the research model in studies involving multiple countries. Going forward, it will help companies that launch m-payment systems internationally to better understand the role of individual- and system-specific characteristics in the acceptance of m-payments.

This paper is organized as follows: Section 2 reviews relevant literature on TRI, TAM, and the integrated technology readiness and acceptance model (TRAM). Section 3 presents the research model and hypotheses; Section 4 explains the research methods. The results are presented and discussed in Section 5, followed by the conclusion in Section 6.

2. Literature Review

2.1. *Technology readiness index*

Anchored in the literature on the adoption of innovations and people–technology interactions, technology readiness (TR) refers to "people's propensity to embrace and use new technologies for accomplishing goals in home life and at work. The construct can be viewed as an overall state of mind resulting from a gestalt of mental enablers and inhibitors that collectively determine a person's predisposition to use new technologies" [Parasuraman, 2000].

In particular, these opinions can be categorized into four dimensions. Optimism (OPT) is defined as "a positive view of technology and a belief that it offers people increased control, flexibility, and efficiency in their lives" [Parasuraman, 2000]. Innovativeness (INN) means that a person has "a tendency to be a technology pioneer and thought leader" [Parasuraman, 2000]. Discomfort (DIS) deals with "a perceived lack of control over technology and a feeling of being overwhelmed by it" [Parasuraman, 2000]. Insecurity (INS) refers to a "distrust of technology, stemming from skepticism about its ability to work properly, and concerns about its potential harmful consequences" [Parasuraman and Colby, 2014].

OPT and INN are drivers of TR, whereas DIS and INS are inhibitors. These positive and negative feelings about technology can coexist, and people can have very different attitudes toward technology [Parasuraman, 2000].

The initial TRI scale consisted of 36 different items. Additionally, Lin and Hsieh [2012] derived a 16-item scale. However, this shortened scale did not include updated or modified items, and it remained a problem that topics identified in the scale were no longer innovative, since there were changes in the technology environment [Parasuraman and Colby, 2014]. Therefore, Parasuraman and Colby [2014] updated and shortened the TRI, which resulted in a new 16-item version — the TRI 2.0. An advantage of this version is that it can be used very well in combination with other constructs and when conducting studies in multiple countries [Parasuraman and Colby, 2014].

2.2. *Technology acceptance model*

Rooted in Fishbein's and Ajzen's Theory of Reasoned Action [Fishbein and Ajzen, 1975; Davis, 1985, 1989] developed the TAM to explain and predict user acceptance of information technology. According to TAM, user acceptance and the intention to use an information system are determined by its Perceived Usefulness (PU) and Perceived Ease of Use (PEoU) [Davis, 1989].

PU is defined as "the degree to which a person believes that using a particular system would enhance his or her job performance" [Davis, 1989]. PEoU, on the other hand, is defined as "the degree to which a person believes that using a particular system would be free of effort" [Davis, 1989].

The definitions of the central variables, PU and PEoU, already hint at the fact that the TAM scales were originally tested in a work environment [Davis, 1989]. However, TAM has also been used in a variety of studies in non-work settings, e.g., in the context of smartphone applications usage [Verkasalo *et al.*, 2010], internet banking [Kesharwani and Bisht, 2012], and acceptance of electronic transactions [Al-Gahtani, 2011].

Additionally, the combination of TAM with various external variables is common practice. Prominent examples include TAM2, which extended TAM by social influence processes and cognitive instrumental processes [Venkatesh and Davis, 2000]; the unified theory of acceptance and use of technology, which unites elements of eight models of individual acceptance [Venkatesh *et al.*, 2003]; and TAM3, which combines TAM2 and the model of determinants of PEoU [Venkatesh, 2000] into a comprehensive model of individual-level information technology adoption and use [Venkatesh and Bala, 2008].

2.3. *Technology readiness and acceptance model*

The TRAM integrates the two aforementioned models, and was presented by Lin *et al.* [2007]. The study stemmed from a questioning of TAM's applicability in non-work settings. TRAM links the TRI with the system-specific dimensions of TAM and thereby investigates how consumers' TR propensities are correlated with their technology acceptance. Results indicate that TRAM increases the applicability and explanatory power of both the prior models in marketing settings [Lin *et al.*, 2007].

Guhr *et al.* [2013] followed this approach and used the TRI as a predictor of the acceptance of m-payment in Finland, Germany, the USA, and Japan. Findings showed that a person's overall TR level had a positive effect on PU and PEoU in all four countries. Hence, TRAM is also applicable in the context of m-payments.

In contrast, other studies [Walczuch *et al.*, 2007; Godoe and Johansen, 2012] measured the relationship between each TRI dimension — OPT, INN, DIS, and INS — and the central dimensions of TAM, which resulted in a more specific model. Both drivers of TR are theorized to lead to higher PU, PEoU [Walczuch *et al.*, 2007], and actual use [Godoe and Johansen, 2012], whereas each inhibitor of TR is thought to be negatively related to the TAM dimensions [Walczuch *et al.*, 2007].

3. Research Model and Hypotheses

Based on the literature review, the research model links each TRI 2.0 variable directly with PU and PEoU, as suggested by Walczuch *et al.* [2007] and replicated by Godoe and Johansen [2012].

The factor "country" is assumed to moderate these relationships. This approach allows for a more thorough comparison of two groups. Following Lin *et al.*'s [2007] research model, Use Intention (UI) is included in this TRAM, since actual m-payment use is not assumed to be common yet and, hence, cannot be presupposed. Direct effects of the TRI 2.0 dimensions on UI are not included in this research model, although Parasuraman [2000] recognized these effects. Furthermore, mediation effects of PU and PEoU will not be tested, but have been proven by Lin *et al.* [2007].

So far, there is little research on the TRI 2.0, given that it was first published in 2014. Thus, the inclusion of UI in a more detailed TRAM, usage of the TRI 2.0 items, measurement invariance testing, and testing of moderating effects of the country of origin of the participants constitute an extension of previously conducted research. Table 1 provides a definition of each research variable. Justifications for the hypotheses, which are largely replicated from Walczuch *et al.* [2007] and the original TAM [Davis, 1989], are presented in the following.

OPT and INN are drivers of TR, and expected to correlate with people's propensities to employ new technologies [Parasuraman, 2000]. People scoring high on these dimensions are more likely to adopt and use high-technology products or services, and they hold positive beliefs about technology [Parasuraman, 2000; Parasuraman and Colby, 2014]. Furthermore, a higher TR level is positively related to PU and PEoU [Lin and Chang, 2011]. Therefore, it is hypothesized that OPT and INN will have a positive effect on PU and PEoU of m-payment.

H1: High personal OPT about technology leads to higher PU of m-payment.

H2: High personal OPT about technology leads to higher PEoU of m-payment.

H3: High personal INN toward technology leads to higher PU of m-payment.

H4: High personal innovativeness toward technology leads to higher PEoU of m-payment.

Table 1. Definitions of the research variables.

Variable	Definition	References
Optimism (OPT)	"A positive view of technology and a belief that it offers people increased control, flexibility, and efficiency in their lives."	Parasuraman [2000]
Innovativeness (INN)	"A tendency to be a technology pioneer and thought leader."	Parasuraman [2000]
Discomfort (DIS)	"A perceived lack of control over technology and a feeling of being overwhelmed by it."	Parasuraman [2000]
Insecurity (INS)	"Distrust of technology, stemming from skepticism about its ability to work properly and concerns about its potential harmful consequences."	Parasuraman and Colby [2014]
Perceived Usefulness (PU)	"The degree to which a person believes that using a particular system would enhance his or her task."	Davis [1989] Gao *et al.* [2014]
Perceived Ease of Use (PEoU)	"The degree to which a person believes that using a particular system would be free of effort."	Davis [1989]
Use Intention (UI)	"A person's subjective probability" that he or she will perform a certain behavior.	Davis [1989] Fishbein and Ajzen [1975] Zarmpou *et al.* [2012]

In contrast, DIS and INS are inhibitors of TR. These two dimensions are associated with a negative attitude toward technology, and people scoring high on DIS and INS are more reluctant to accept technology [Parasuraman, 2000; Tsikriktsis, 2004]. Therefore, it is hypothesized that DIS and INS will have a negative effect on the dimensions of TAM. The only exceptions are DIS and PU, which are not expected to be negatively or positively related. It is assumed that a system can still be seen as useful, even if the person is not comfortable with using technology in general [Walczuch *et al.*, 2007; Godoe and Johansen, 2012].

H1: High personal DIS in using technology does not lead to higher or lower PU of m-payment.
H2: High personal DIS with regard to technology leads to lower PEoU of m-payment.
H3: High personal INS with regard to technology leads to lower PU of m-payment.
H4: High personal INS of technology leads to lower PEoU of m-payment.

There is a sound basis of empirical evidence that PU and PEoU are determinants of the intention to use and the actual use of a system [Venkatesh and Davis, 2000; Davis *et al.*, 1989; Adams *et al.*, 1992]. PEoU is both directly and indirectly linked to UI through its impact on PU [Venkatesh and Davis, 2000; Davis, 1989]. PU also strongly influences the intention to use a system and acts as the major determinant of self-reported current usage and self-predicted future usage [Davis, 1989; Davis *et al.*, 1989]. As such, the relationships hypothesized by the majority of TAM studies are replicated.

H1: PEoU of m-payment is positively related to PU of m-payment.
H2: PEoU of m-payment is positively related to the intention to use m-payment.
H3: PU of m-payment is positively related to the intention to use m-payment.

Since this study does not directly measure cultural dimensions, it is not possible to investigate whether a link between cultural factors and m-payment adoption can be established [Straub *et al.*, 1997]. But given that cultural differences are believed to exist because Germany and South Africa differ on the scores they achieve on Hofstede's cultural dimensions [Hofstede, 2001], one of the most commonly cited indexes, and given that the m-payment infrastructure and people's readiness to adopt the technology differs [Mastercard, 2012; Berndt *et al.*, 2010], it is assumed that the factor "country" has a moderating effect on the above-mentioned structural relationships of TRAM.

H1: The factor "country" moderates the structural relationships between OPT, INN, INS, DIS, PEoU, PU, and UI.

Figure 1 presents the research model, including the hypothesized relationships.

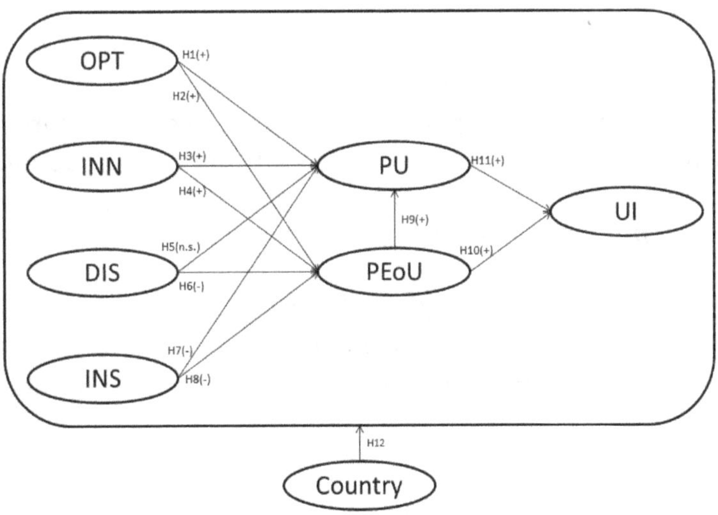

Figure 1. Research model with hypothesized relationships between variables.

4. Research Methods

The constructs were derived from the literature review in Section 2. The first part of the questionnaire replicates the items of TRI 2.0 [Parasuraman and Colby, 2014] and, hence, consists of 16 statements whose order was randomized as recommended by Parasuraman and Colby [2014].[a] The items listed in the second half were adapted from Davis [1989] and UI was adapted from Zarmpou *et al.* [2012] to fit into the context of m-payments. PU and PEoU each consist of six items, whereas there are four UI items. Each item was rated on a 5-point Likert scale coded as 1 = strongly disagree, 2 = somewhat disagree, 3 = neutral, 4 = somewhat agree, and 5 = strongly agree.

The following definition of m-payment was provided to the participants to ensure that all of them had a common understanding of the technology: "Mobile payments, for the purposes of this research, are defined as a payment procedure whereby the payer uses mobile electronic devices (such as smartphones or tablets) to make a payment."

The questionnaire was translated into German by a native speaker, the principal researcher. A pretest with students from each country revealed some translation errors in the German version, as well as discrepancies between the original American English version and the South African manner of speaking. These issues were discussed and resolved together with the pretest participants and academic advisors from each university.

The survey was conducted with a sample consisting of 253 participants from a German university and 200 participants from a South African university. The student samples are

[a]These questions comprise the TRI 2.0 which is copyrighted by Parasuraman and Rockbridge Associates, Inc., 2014. This scale may be duplicated only with written permission from the authors.

assumed to be alike with respect to demographic characteristics. This helps reduce bias [van Herk *et al.*, 2005]. In academic studies, using matched samples is common practice [Schaffer and Riordan, 2003]. The questionnaire was administered to the students during their lectures to ensure a high response rate.

The research model was tested using structural equation modeling in IBM SPSS Amos 23. The covariance-based approach is applied since it has proved to be very robust with regard to violations of its underlying distributional assumptions [Reinartz *et al.*, 2009].

5. Results and Discussion

5.1. *Preliminary analyses and sample description*

During the missing values analysis, participants who had answered less than 90% of the questionnaire were excluded from subsequent analyses, which resulted in a final sample size of 444 participants. The sample was divided between people who answered the questionnaire at the German university and those who answered the questionnaire at the South African university. The South African sample consisted of 195 participants and the German sample, 249 participants. Checks for unengaged responses and outliers did not reveal the necessity to exclude further cases. Data that was missing at random and relevant for research variables was imputed using the expectation-maximization algorithm in IBM SPSS Statistics 22 [Dempster *et al.*, 1977]. The demographic characteristics of the sample are presented in Table 2.

A preliminary analysis was conducted and revealed no violation of the assumptions of multicollinearity, specifically, none of the predictor variables were significantly correlated with one another.

Table 2. Demographic characteristics of the sample.

	South Africa	Germany		South Africa	Germany
	Frequency/%	Frequency/%		Frequency/%	Frequency/%
Gender			Have used m-payment before		
Male	75/38.5	88/35.3	Yes	117/60.0	74/29.7
Female	120/61.5	161/64.7	No	71/36.4	174/69.9
No reply	0/0.0	0.0/0.0	No reply	7/3.6	1/0.4
Total	195/100.0	249/100.0	Total	195/100.0	249/100.0
Age			Nationality		
18–23 years	129/66.2	158/63.5	South African	98/50.3	0/0.0
24–29 years	47/24.1	79/31.7	German	0/0.0	207/83.1
30 years and above	16/8.2	8/3.2	Other	52/26.7	9/3.6
No reply	3/1.5	4/1.6	No reply	45/23.0	33/13.3
Total	195/100.0	249/100.0	Total	195/100.0	249/100.0

5.2. *Model fit*

The baseline models were established with the whole dataset consisting of both German and South African participants. Measurement models were tested separately in order to detect flaws faster.

TAM had initially been below the general cut-off criteria for most fit indexes, as shown in Table 3. Therefore, modification indices were inspected and three error covariances were added to achieve an acceptable fit. The limit for low factor loadings was set at 0.50 [Janssens *et al.*, 2008]. All TAM items exceeded this value and, hence, there was no need to exclude items.

TRI could not be estimated by Amos, because of a non-positive definite sample covariance matrix. The item OPT2 was detected to inhibit the estimation and, therefore, was excluded from subsequent analyses. The estimation of TRI revealed that the measurement model lacked sufficient model fit. In addition, nine out of 15 factor loadings were below the general threshold of 0.50. However, deleting these items would have led to an unidentified model. Therefore, only the item with the lowest loading was deleted (INS1). Nevertheless, TRI still exhibited a weak model fit and a not positive definite covariance matrix; this could not be resolved.

The adjusted versions of both measurement models were combined to the structural model TRAM and covariance arrows

Table 3. Model fit (for all data together).

	χ^2/df	GFI	AGFI	TLI	CFI	RMSEA
Measurement models						
TAM (initial)	4.192	0.89	0.852	0.924	0.936	0.085
TAM (adjusted)	2.664	0.932	0.906	0.96	0.968	0.061
TRI (initial)	2.665	0.936	0.909	0.739	0.791	0.061
TRI (adjusted)	2.492	0.945	0.919	0.781	0.829	0.058
Structural model						
TRAM (initial)	2.107	0.887	0.863	0.919	0.928	0.05

were removed. An additional constraint was placed on the error term of the item INS4 to avoid a not positive definite covariance matrix. TRAM had an acceptable model fit at best, with values close to the cut-off criteria $\chi^2/df \leq 2.5$, GFI ≥ 0.9, AGFI ≥ 0.9, CFI ≥ 0.9, RMSEA ≤ 0.05 [Homburg and Baumgartner, 1995], and TLI ≥ 0.95 [Hu and Bentler, 1999].

5.3. *Hypotheses testing*

The hypotheses were tested using multiple group analysis in Amos. The results show that not all of the TRI 2.0 dimensions had a significant influence on the TAM variables, although only the relationship between DIS and PU was expected to be insignificant. The standardized beta coefficients were, however, in the predicted directions.

An inspection of the estimation results for the South African sample revealed that only OPT had a significant effect on PU and only INN had a significant influence on PEoU, in the expected directions. Thus, H1 and H4 were supported, whereas H2, H3, H6, H7, and H8 were rejected. H5 was supported because the relationship between DIS and PU was expected to be not significant. In addition, all hypotheses related to the TAM were supported (H9, H10, and H11). PU was the strongest predictor of UI ($\beta = 0.558$, t-value $= 5.788$, p-value < 0.001).

Similarly, H1, H4, and H5 were supported for the German sample. That is to say, OPT had a significant influence on PU and INN had a significant influence on PEoU. Furthermore, DIS — as expected — did not have a significant influence on PU.

Additionally, H6, H7, and H8 were supported. Thus, in the German dataset, DIS had a significant influence on PEoU and INS had a significant influence on PU and PEoU. For the TAM part of the model, however, only H9 and H11 were supported, whereas PEoU did not have a significant influence on the intention to use m-payment (i.e., H10 was rejected). Again, PU was the strongest predictor of UI ($\beta = 0.777$, t-value $= 11.307$, p-value < 0.001).

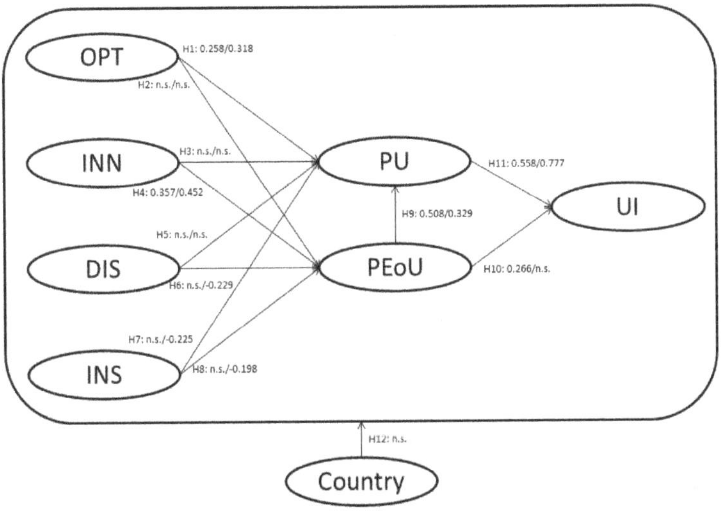

Figure 2. Research model and estimation results (South Africa/Germany).
Note: n.s.= not significant.

This implies that, in both samples, participants who are optimistic with regard to new technologies are more likely to find mobile payments useful, whereas people with innovative personality characteristics will find systems easy to use. The inhibitor dimensions DIS and INS, however, do not have a significant influence on the PU and PEoU of mobile payments in South Africa (see Figure 2). In the South African sample, PEoU of mobile payments has an influence on PU and UI. These results are in accordance with the majority of previously conducted TAM-related studies.

A significant relationship between the inhibitors and the system-specific dimensions was, indeed, found in the German sample, indicating that people who are insecure and not comfortable with new technologies will find mobile payments less useful and less easy to use. Surprisingly, PEoU did not have a significant influence on the intention to use mobile payments. This contradicts most previous TRAM-related studies.

Table 4 provides a detailed summary of estimation results for both samples.

Table 4. Summary: Estimation results of the structural model (standardized estimates).

		South Africa				Germany			
Hypothesis	Relationship	β	*t*-value	*p*-value	Conclusion	β	*t*-value	*p*-value	Conclusion
H1	OPT → PU	0.258	2.359	0.018	supported	0.318	3.156	0.002	supported
H2	OPT → PEoU	0.096	0.926	0.354	n.s./rejected	0.073	0.887	0.375	n.s./rejected
H3	INN → PU	0.102	1.025	0.305	n.s./rejected	0.051	0.598	0.55	n.s./rejected
H4	INN → PEoU	0.357	3.197	0.001	supported	0.452	4.962	***	supported
H5	DIS → PU	0.19	1.363	0.173	supported	0.122	1.396	0.163	supported
H6	DIS → PEoU	–0.204	–1.389	0.165	n.s./rejected	–0.229	–2.481	0.013	supported
H7	INS → PU	–0.167	–1.726	0.084	n.s./rejected	–0.225	–2.57	0.01	supported
H8	INS → PEoU	–0.052	–0.498	0.618	n.s./rejected	–0.198	–2.35	0.019	supported
H9	PEoU → PU	0.508	4.798	***	supported	0.329	3.629	***	supported
H10	PEoU → UI	0.266	3.038	0.002	supported	0.073	1.374	0.17	n.s./rejected
H11	PU → UI	0.558	5.788	***	supported	0.777	11.307	***	supported
H12	The factor "country" as a moderator	ΔCFI < 0.01 (if two measurement weights are set free). This indicates invariance of the structural model; thus, H12 must be rejected.							

Notes: *** Significance level < 0.001; n.s. = not significant; Model Fit: South Africa: $\chi^2/df = 1.545$, GFI = 0.829, AGFI = 0.797, TLI = 0.846, CFI = 0.861, RMSEA = 0.053; Germany: $c^2/df = 2.061$, GFI = 0.818, AGFI = 0.784, TLI = 0.875, CFI = 0.887, RMSEA = 0.065.

Next, the moderating effect of the factor "country", H12, was also tested using multiple group analysis in Amos. The unconstrained model was used as a baseline model for subsequent model comparisons. It achieved an acceptable model fit at most (χ^2/df = 1.803, GFI = 0.823, AGFI = 0.79, TLI = 0.866, CFI = 0.879, RMSEA = 0.043).

Constraints were then imposed on measurement weights and structural weights. The CFI test for the measurement weights model revealed that ΔCFI was slightly too high (ΔCFI = 0.011). Therefore, two measurement weights were set free to achieve partial invariance.

For the structural weights model, which was critical to testing H12, the invariant measurement weights and all structural weights were constrained to be equal across groups. ΔCFI was equal to 0.0 in comparison with the partial measurement weights model, and equal to 0.01 in comparison with the unconstrained model. Hence, the paths were not significantly different between the two datasets, and H12 was rejected. The multiple group analysis also revealed the invariance of even stricter constrained models, as shown in Table 5.

Table 5. Structural invariance: Model comparisons.

	Invariance assessment	
Structural model TRAM	CFI	ΔCFI
Unconstrained	0.879	n.a.
Measurement weights (initial)	0.868	0.011
Measurement weights (partial)	0.869	0.01
Structural weights	0.869	0.0
Structural covariances	0.869	0.0
Structural residuals	0.863	0.006
Measurement residuals	0.769	0.094

Note: n.a. = not applicable.

6. Conclusion

This study investigated the relationships between TRI 2.0 dimensions and the variables of the TAM. Moreover, it contributed to TR and acceptance research by using the latest TRI 2.0 items and testing whether the integrated research model TRAM applies to Germany and South Africa. To this end, a survey was conducted at one university each in Germany and South Africa.

The results show that the measurement models achieved partial scalar invariance. The structural model was invariant across countries and had an acceptable model fit for the data. Some, but not all, TRI 2.0 dimensions had a significant influence on the PU and PEoU of m-payment. In both countries, PU had the strongest influence on the intention to use m-payment and OPT had the strongest influence on PU.

Apart from its contribution to TR and acceptance research, the results of this study are also relevant for practitioners in the fields of developing and marketing m-payment systems.

In the development phase of m-payment systems, for instance, during the development of applications that enable users to make a payment with their mobile devices, companies should focus on those functions that are the most beneficial to the end user. At the same time, they should pay attention to the user experience, i.e., the ease of use of a system, since this, in turn, influences its PU. This can, for example, apply to the m-payment smartphone application that a company provides. The user interface and user experience of such an app should be intuitive, granting users an overview and easy access to all of its functions, so that they can quickly grasp its value to them. This is also supported by a recent, representative BITKOM [2016] publication, stating that although one-third of German smartphone users have a positive attitude toward mobile payments, the technology is only sparsely adopted because of a poor customer education.

When marketing the product (for example the m-payment app), managers should be aware of their target group's personality characteristics and TR, as this will influence adoption choices. Moreover, when doing business in international markets, it is crucial to know whether differences exist with regard to the personality characteristics that actually influence the adoption process. Hence, the marketing strategy must be tailored specifically to the target group and its TR profile. An example that emphasizes the importance of a system's PU and of a marketing strategy that matches the target group, is M-Pesa in Kenya and other infrastructure-constrained countries. The mobile phone-based financial service enables users to transfer money, in addition to other use cases, which is a clear advantage for many workers that regularly send money home to family members in rural areas. These benefits, amongst others, were clearly communicated in a marketing campaign [The Economist, 2013].

It should be mentioned that some limitations exist with regard to this study. First, the questionnaire items were translated for the German survey and adapted for the South African survey. In addition, many students from the South African sample were not native speakers of English. Another approach to translation, for instance, a combination of back translation (for the German questionnaire) and the committee approach (for the South African questionnaire), might lead to better results [van Herk *et al.*, 2005].

Second, the sample might have been subject to bias because it consisted only of students and, hence, was not representative of the total German or South African population.

Third, not all the participants had used m-payment before and, therefore, might have experienced difficulties in answering the TAM questions. In addition, a very broad definition of m-payment was given to the participants. It would be more precise to narrow down the definition, for example, to a specific m-payment technology. An overview of electronic and mobile payment systems and technology differences between developed and developing countries is provided by Mathew *et al.* [2014].

Fourth, cultural dimensions were not measured directly, nor were other circumstances, such as infrastructure, taken into account by the research model. Therefore, it is not possible to say that a link between these factors and m-payment adoption has been empirically established [Straub *et al.*, 1997].

And fifth, the TRI 2.0 measurement model lacked a good model fit, convergent validity, reliability, and discriminant validity. Parasuraman and Colby [2014] also acknowledge that the scales for DIS and INS show weak performances on some psychometric criteria and suggest that researchers should investigate if this can be resolved, for instance, through refined wording.

In addition to refining the TRI 2.0 scales for surveys with non-native English speakers, there are more opportunities for future research. Examining the structural stability and invariance of TRAM across more and other countries, as well as other technologies, is one potential way. Replicating this study in Germany and South Africa, but with a larger, representative sample, in order to investigate whether the model performs better on psychometric criteria, is another research avenue.

For future research, it would also be interesting to measure or include cultural dimensions in the research model. Using Hofstede's cultural dimensions as moderators, analogous to the work of Hung and Chou [2014], would enable researchers to identify cultural influences on the adoption of m-payment.

Another idea would be to investigate if the banking infrastructure plays a role in consumer adoption of m-payment and if there is a difference between developed and developing countries. This idea stems from the fact that, on a percentage basis, more of the South African students had used m-payment earlier, and the country was assumed to have a less-developed banking infrastructure.

Ultimately, as PU had the strongest influence on the intention to use m-payment, researchers could also investigate which system-specific characteristics benefit a high usefulness of m-payment, and include these in an extended TAM.

References

Adams, D. A., Nelson, R. R. and Todd, P. A. (1992). Perceived usefulness, ease of use, and usage of information technology: A replication. *MIS Quarterly*, **16**(2):227–247.

Agarwal, R. and Prasad, J. (1999). Are individual differences germane to the acceptance of new information technologies? *Decision Sciences*, **30**(2):361–391.

Al-Gahtani, S. S. (2011). Modeling the electronic transactions acceptance using an extended technology acceptance model. *Applied Computing and Informatics*, **9**(1):47–77.

Berndt, A. D., Saunders, S. G. and Petzer, D. J. (2010). Readiness for banking technologies in developing countries. *Southern African Business Review*, **14**(3):47–76.

BITKOM (2014). Leitfaden Mobile Wallet. Referenced on Jan 03, 2015, http://www.bitkom.org/de/publikationen/38337_80723.aspx.

BITKOM (2016). Bezahlen mit dem Smartphone funktioniert, aber kaum jemand weiß wie. Reference on Jul 31, 2016, https://www.bitkom.org/Presse/Presseinformation/Bezahlen-mit-dem-Smartphone-funktioniert-aber-kaum-jemand-weiss-wie.html.

Chen, L.-d. (2008). A model of consumer acceptance of mobile payment. *International Journal of Mobile Communications*, **6**(1): 32–52.

Davis, F. D. (1985). A Technology Acceptance Model for empirically testing new end-user information systems: Theory and results. Doctoral Dissertation, MIT Sloan School of Management, Cambridge, MA.

Davis, F. D. (1989). Perceived usefulness, perceived ease of use, and user acceptance of information technology. *MIS Quarterly*, **13**(3): 319–340.

Davis, F. D., Bagozzi, R. P. and Warshaw, P. R. (1989). User acceptance of computer technology: A comparison of two theoretical models. *Management Science*, **35**(8):982–1003.

Dempster, A. P., Laird, N. M. and Rubin, D. B. (1977). Maximum likelihood from incomplete data via the EM algorithm. *Journal of the Royal Statistical Society, Series B (Methodological)*, **39**(1):1–38.

Erumban, A. A. and de Jong, S. B. (2006). Cross-country differences in ICT adoption: A consequence of culture? *Journal of World Business*, **41**:302–314.

Fishbein, M. and Ajzen, I. (1975). Belief, attitude, intention and behavior: An introduction to theory and research. Addison-Wesley, Reading, MA.

Gao, S., Krogstie, J. and Siau, K. (2014). Adoption of mobile information services: An empirical study. *Mobile Information Systems*, **10**: 147–171.

Godoe, P. and Johansen, T. S. (2012). Understanding adoption of new technologies: Technology readiness and technology acceptance as an integrated concept. *Journal of European Psychology Students*, **3**:38–52.

Guhr, N., Loi, T., Wiegard, R. and Breitner, M. H. (2013). Technology readiness in customers' perception and acceptance of m(obile)-payment: An empirical study in Finland, Germany, the USA, and Japan. *11th International Conference on Wirtschaftsinformatik*, 27th February — 01st March 2013, Leipzig, Germany, pp. 119–133.

Hofstede, G. (2001). *Culture's Consequences: Comparing Values, Behaviors, Institutions, and Organizations across Nations*. Sage Publications Ltd.

Homburg, C. and Baumgartner, H. (1995). Beurteilung von Kausalmodellen: Bestandsaufnahme und Anwendungsempfehlungen. *Marketing: Zeitschrift für Forschung und Praxis*, **17**(3): 162–176.

Hu, L.-T. and Bentler, P. M. (1999). Cutoff criteria for fit indexes in covariance structure analysis: Conventional criteria versus new alternatives. *Structural Equation Modeling: A Multidisciplinary Journal*, **6**(1): 1–55.

Hung, C.-L. and Chou, J. C.-L. (2014). Examining the cultural moderation on the acceptance of mobile commerce. *International Journal of Innovation and Technology Management*, **11**(2):1450010.

Janssens, W., Wijnen, K., De Pelsmacker, P. and Van Kenhove, P. (2008). *Marketing Research with SPSS*. Pearson Education Limited.

Kesharwani, A. and Bisht, S. S. (2012). The impact of trust and perceived risk on internet banking adoption in India: An extension of technology acceptance model. *International Journal of Bank Marketing*, **30**(4):303–322.

Lin, J.-S. C. and Chang, H.-C. (2011). The role of technology readiness in self-service technology acceptance. *Managing Service Quality: An International Journal*, **21**(4):424–444.

Lin, J.-S. C. and Hsieh, P.-L. (2012). Refinement of the technology readiness index scale: A replication and cross-validation in the self-service technology context. *Journal of Service Management*, **23**(1):34–53.

Lin, C.-H., Shih, H.-Y. and Sher, P. J. (2007). Integrating technology readiness into technology acceptance: The TRAM model. *Psychology & Marketing*, **24**(7):641–657.

Luarn, P. and Lin, H.-H. (2005). Toward an understanding of the behavioral intention to use mobile banking. *Computers in Human Behavior*, **21**:873–891.

MasterCard (2012). The mobile payments readiness index: A global market assessment. Whitepaper, Reports about Germany and South Africa, Referenced on Feb 3, 2014, http://mobilereadiness.mastercard.com/reports/.

Mathew, M., Magnier-Watanabe, R., Pratheeba, S. and Balakrishnan, N. (2014). Assessing technology differences in electronic and mobile payment systems among developed and developing countries. *International Journal of Innovation and Technology Management*, **11**(2):1450005.

Parasuraman, A. (2000). Technology readiness index (Tri): A multiple-item scale to measure readiness to embrace new technologies. *Journal of Service Research*, **2**:307–320.

Parasuraman, A. and Colby, C. L. (2014). An updated and streamlined technology readiness index: TRI 2.0. *Journal of Service Research*, **18**(1):59–74.

Reinartz, W. J., Haenlein, M. and Henseler, J. (2009). An empirical comparison of the efficacy of covariance-based and variance-based SEM. INSEAD Working Paper No. 2009/44/MKT, Referenced on Mar 30, 2015, SSRN: http://ssrn.com/abstract=1462666.

Schaffer, B. S. and Riordan, C. M. (2003). A review of cross-cultural methodologies for organizational research: A best-practices approach. *Organizational Research Methods*, **6**(2):169–215.

Singh, S. (2006). Cultural differences in, and influences on, consumers' propensity to adopt innovations. *International Marketing Review*, **23**(2):173–191.

Straub, D., Keil, M. and Brenner, W. (1997). Testing the technology acceptance model across cultures: A three country study. *Information & Management*, **33**:1–11.

The Economist (2013). Why does Kenya lead the world in mobile money? Referenced on Jul 31, 2016, http://www.economist.com/blogs/economist-explains/2013/05/economist-explains-18.

Tsikriktsis, N. (2004). A technology readiness-based taxonomy of customers. A replication and extension. *Journal of Service Research*, **7**(1):42–52.

van Herk, H., Poortinga, Y. H. and Verhallen, T. M. M. (2005). Equivalence of survey data: Relevance for international marketing. *European Journal of Marketing*, **39**(3):351–364.

Venkatesh, V. (2000). Determinants of perceived ease of use: Integrating control, intrinsic motivation, and emotion into the Technology Acceptance Model. *Information Systems Research*, **11**(4):342–365.

Venkatesh, V. and Bala, H. (2008). Technology Acceptance Model 3 and a research agenda on interventions. *Decision Sciences*, **39**(2): 273–315.

Venkatesh, V. and Davis, F. D. (2000). A theoretical extension of the Technology Acceptance Model: Four longitudinal field studies. *Management Science*, **46**(2):186–204.

Venkatesh, V., Morris, M. G., Davis, G. B. and Davis, F. D. (2003). User acceptance of information technology: Toward a unified view. *MIS Quarterly*, **27**(3):425–478.

Verkasalo, H., López-Nicolás, C., Molina-Castillo, F. J. and Bouwman, H. (2010). Analysis of users and non-users of smartphone applications. *Telematics and Informatics*, **27**(3):242–255.

Walczuch, R., Lemmink, J. and Streukens, S. (2007). The effect of service employees' technology readiness on technology acceptance. *Information & Management*, **44**:206–215.

Zarmpou, T., Saprikis, V., Markos, A. and Vlachopoulou, M. (2012). Modeling users' acceptance of mobile services. *Electronic Commerce Research*, **12**:225–248.

Chapter 8

Emerging Applications: Mobile Food Information

Numtip Trakulmaykee, Thakerng Wongsirichot, and Yaowalak Trakulmaykee

1. Introduction

The number of mobile application users has continually grown over the last decade. According to Flurry Insights [2015], overall mobile application usage grew by 76% in the year 2014. Mobile-based researchers and businesses still study the factors influencing the acceptance of application technologies in different user groups today. Technological acceptance among different user groups may be influenced by different factors. Tsang *et al.* [2011] looked at the needs of guidebook users and found differences in security needs and functional needs between two types of travelers: independent tourists and comprehensive package tourists. Therefore, the factors influencing international tourists' uptake of mobile food information (MFI) may be different between groups of the independent tourists and package tourists in this study.

Most previous studies also focused on the dependence of users' acceptance on particular technologies and user groups. Feng *et al.* [2015] studied mobile tourist guides built on location-based technology and cloud platforms. Their study focused on the technological qualities of stability, flexibility, security and scalability. Meanwhile, a study on mobile shopping [Agrebi and Jallais, 2015] focused on three factors: perceived usefulness, perceived ease-of-use, and perceived enjoyment. In general, international tourists will spend more than three days traveling and need information for their meals during trips, especially about the ingredients. They need to know about ingredients and cooking methods for reasons of food allergies, religion, faith, and health concerns. For example, Islamic tourists cannot eat pork, Hindu tourists cannot eat beef, vegetarian tourists can eat only vegetables, and food-allergic tourists cannot eat some kinds of food such as pepper, peanuts, garlic, seafood, cheese, nuts, spicy food, milk, and diary. Moreover, most menus of local food are cooked with different ingredients from the international tourist's usual food. Therefore, in the context of MFI, international tourists may judge the quality of information by its perceived usefulness and perceived ease-of-use. Furthermore, other factors could influence usage; interactivity, presentation, perceived compatibility, and perceived mobility could all play a part.

In the context of MFI, there has been a lack of investigation of mobile design qualities related to information, interaction, and presentation, and also of tourists' perceptions of usefulness, ease-of-use, compatibility, and mobility. In particular, there is little research comparing the influences of these factors on uptake among two international tourist types: the independent tourist and the package tourist. Therefore, this study aims to investigate these factors and to compare their influences on the intention to use MFI, for both independent tourists and package tourists. In this study, the MFI refers to a mobile application for international tourists. Information about food and recommended restaurants is provided, based on dietary requirements (e.g., religion, faith, food allergy, and health concern). Examples of the food information in MFI are the top 10 local food menus, the ingredients of those menus, and recommended menus based on tourists' dietary requirements. For the recommended restaurant information, the MFI provides information about nearby restaurants, popular restaurants, directions, and locations, and recommends restaurants based on tourists' dietary requirements.

2. Intention to Use Information Technology (IT) in Tourism

In general, any episode of travel consists of three phases based on time and the tourist's activities; these phases can be described as before, during, and after the trip [Chung and Buhalis, 2008; Simon-Wong and Gladys-Liu, 2011; Tsang *et al.*, 2011] as presented in Table 1.

Table 1. Tour information involvement in travel phases.

Travel phase	Tour information
Phase 1: Before the trip	Tour information is potentially the greatest factor in planning and making decisions for tourists in their travels [Jenkin, 2010]. In terms of travel planning, tourists want to know information about culture, transportation, hotels, how to get there, maps, local food, restaurants, prices, activities, and attractions [Tsang *et al.*, 2011; Yu and Chang, 2009]. Information from previous travelers' experience is very useful for decision-making because tourists are able to find information easily and quickly in many types of content such as photos, texts, and videos.
Phase 2: During the trip	In one of many possible scenarios, the tourists desire to get information while they are on the road or during their travel [(Chung and Buhalis (2008); Tsang *et al.*, (2011); Yu and Chang (2009)]. The information which tourists need during their trip is about restaurants, attractions, emergency phone numbers, transportation, souvenir shops, and hotels. In another scenario, a tourist enjoys taking photos during travel [Luo *et al.*, 2010] and enjoys sharing these photos and videos with other people [Buriano, 2006]. Social networks are a very popular channel for sharing travel experiences, travel impressions, and photos with others.

(*Continued*)

Table 1. (*Continued*)

Travel phase	Tour information
Phase 3: After the trip	Most tourists want to share their experiences with friends, family, or unknown people face-to-face or via social networks [Tsang *et al.*, 2011]. Today, the popular social networks for sharing trip experiences are Facebook, Blog, Twitter, and the mobile application line. With social networks, the tourist can share his or her trip experiences, impressions, and comments with others by posting text, uploading photos, and sharing videoclips [Chung and Buhalis, 2008]. Popular categories of shared content are food menus, restaurants, accommodation, transportation, security, and places of interest.

Source: Compilation by authors.

In the mobile IT context, tour information can enhance an international tourist's travel experiences. It allows tourists to seek relevant travel information through existing technologies such as smart phones, tablets, and phablets. In addition, sharing experiences with others anytime and anywhere also increases tourists' travel knowledge [Peres *et al.*, 2011; Tsang *et al.*, 2011; Yu and Chang, 2009]. Previous studies suggested that the tourist guide should provide tour information based on tourists' needs. Therefore, it is necessary to study the factors affecting international tourists' intention to use MFI, because many tourists need food information which responds to their dietary needs.

3. Technology Acceptance Model

Previous scholarly contributions have typically referred to the technology acceptance model (TAM) as a conceptual framework to identify users' intention to use IT, and it is a popular model to explain IT acceptance variance among users [Ayeh *et al.*, 2013; Chang and Pan, 2011; Gerpott, 2011; Pham and Ho, 2015; Xue *et al.*, 2012]. The TAM was originally developed to improve the understanding of IT usage among employees in organizational contexts [Davis, 1989]. It was designed to measure how perceived usefulness and perceived ease-of-use affect a user's attitude. Davis [1989] demonstrated two important associations; one between a user's attitude and behavioral intention to use technology, and another between behavioral intention and actual use. Although TAM is purportedly used to explicate technology adoption, the constructs of the model are actually meant to be general and universal. Most researchers used TAM and integrated the identified factors to formulate their proposed model and then found that perceived usefulness and perceived ease-of-use were the important factors affecting users' intention to use IT [Ayeh *et al.*, 2013; Phan and Daim, 2011; Xue *et al.*, 2012]. They suggested that more attractive factors should be added to reach a more comprehensive understanding of the additional factors in the users' intention to use mobile IT and IT in general.

4. Proposed Model and Hypotheses

In this study, the proposed model has two main factor categories. The first category is tourist's perceptions that consist of four dimensions: perceived usefulness, perceived ease-of-use, perceived compatibility, and perceived mobility. For the second category, there are three quality dimensions of mobile design: information quality, interaction quality, and presentation quality.

4.1. *Tourist's perceptions*

The proposed development model of this paper (see Figure 1) was based on TAM [Davis, 1989]. The TAM has successfully proven the significance of two factors of user's intention to use technology: perceived usefulness and perceived ease-of-use.

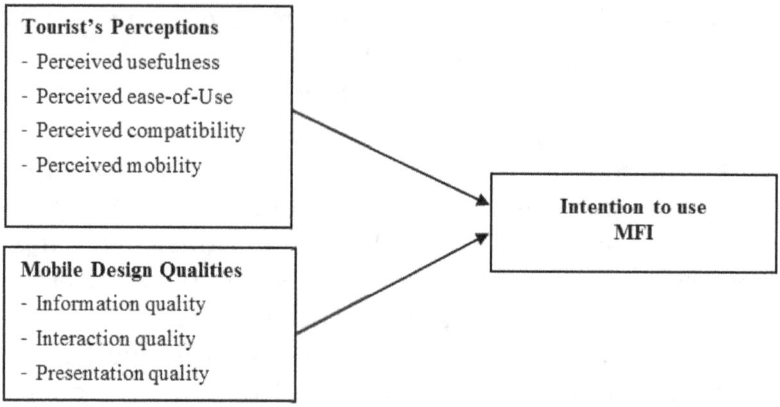

Figure 1. Proposed model.

However, it is possible that there are more factors in each different context which affect a user's needs of the technology. When considering tourists' behavior and mobile features, the perceived compatibility and perceived mobility are possible factors in international tourists' intention to use MFI. Therefore, the tourist's perceptions of MFI should consist of perceived usefulness, perceived ease-of-use, perceived compatibility, and perceived mobility. This study focuses on these four perceptions as the predictors or independent variables.

Perceived usefulness is an important factor directly influencing the users' willingness to use technology [Davis, 1989]. It is often emphasized as a successful key to achieve users' intention to use mobile IT. Kim *et al.* [2008] demonstrated that the perceived usefulness was a key element in attracting travelers to use mobile technologies. It was defined as the degree to which a person believes that using a particular system would enhance his/her performance in TAM. In this study, perceived usefulness is defined as the degree to which any international tourist believes that using the MFI application would enhance his/her travel experience. Most IT studies found that perceived usefulness positively impacted on users' intentions [Ayeh *et al.*, 2013; Chang and Pan, 2011; Gerpott, 2011; Kim and Shin, 2015;

Yadav *et al.*, 2016], hence there should be a positive relationship between perceived usefulness and the intention to use MFI. The hypothesis is proposed as follows:

H1: Perceived usefulness has a positive impact on intention to use MFI.

Perceived ease-of-use is a major factor of technological acceptance [Davis, 1989]. Several previous studies mentioned the important role of perceived ease-of-use in their works. According to Rogers [2003], complexity in an innovation can lead to user resistance due to a lack of user skills and knowledge [Rogers, 2003]. Lawson-Body *et al.* [2012] mentioned that more complex the innovation will be, the lower its rate of adoption, especially for technology on mobile devices. The use of mobile devices requires additional cognitive and mental efforts because of the limited input capability and small screen [Rukzio *et al.*, 2008]. The perceived ease-of-use in TAM [Davis, 1989] is defined as the degree to which a person believes that using a particular system would be free of effort. In the MFI context, the perceived ease-of-use is defined as the degree of ease associated with understanding and usage. Most empirical studies found a relationship between perceived ease-of-use and intention to use technologies [Brown *et al.*, 2003; Kim and Shin, 2015; Kim *et al.*, 2010; Xue *et al.*, 2012], although some previous studies did not find an association between them [Gerpott, 2011; Jung *et al.*, 2012; Tan and Teo, 2000]. This study believes that perceived ease-of-use has a possible impact on tourists' intentions to use MFI. Therefore, the following hypothesis is postulated:

H2: Perceived ease-of-use has a positive impact on intention to use MFI.

Nowadays, technological devices are developed to support the work and lifestyle of users. Compatibility has been one of the most important factors affecting users' intention to use technologies [Rogers, 2003]. In general, high incompatibility will adversely affect potential adopters of an innovation which decreases the likelihood of adoption [Hill and Troshani, 2007]. Users will be unwilling to accept a new technology if that technology is not compatible with their work [Kuo and Lee, 2011]. In the IDT [Rogers, 2003], *perceived compatibility* is generally defined as the degree to which an innovation is perceived to be consistent with the existing values, past experiences, and needs of potential adopters. *Perceived compatibility* in this study is defined as the degree to which an innovation is perceived as being consistent with the user's existing values, lifestyle, and past experience of travel. Most previous studies found empirical evidence to support the role of compatibility as an important factor on intention to use technology. Gerpott [2011], Schierz *et al.* [2010], and Xue *et al.* [2012], even Dai and Palvi [2009], found compatibility did not impact the intention to use mobile commerce in China. This study believes a higher perceived compatibility will result in a greater intention to use MFI. Hence, the hypothesis is proposed as follows:

H3: Perceived compatibility has a positive impact on intention to use MFI.

Mobility is a common feature of mobile devices. Generally, most mobile users decide to buy mobile devices for their mobility and convenience. Schierz *et al.* [2010] accounted mobility as key to users' adoption in the context of mobile services. In the context of MFI for international tourists, perceived mobility is considered as a factor of international tourists' intention to use MFI because international tourists are always moving around during their travel. *Perceived mobility* is defined as the number of benefits provided by mobile technologies. Benefits are considered to be information access, communication, and services anytime and anywhere [Kim *et al.*, 2010]. In addition, Schierz *et al.* [2010] defined individual mobility as the degree to which an individual pursues a mobile lifestyle. In this study, *perceived mobility* is defined as the degree of access to tour information and services anytime and anywhere provided by MFI. Most studies found mobility to be key to users' mobile service adoption, and influenced intention to use technology [Kaba and Osei-Bryson, 2009; Schierz *et al.*, 2010]. Thus, this study expects that the higher perceived mobility of MFI will increase international tourists' intention to use MFI. The hypothesis is proposed as follows:

H4: Perceived mobility has a positive impact on intention to use MFI.

4.2. *Mobile design qualities*

Mobile devices are small handheld computing devices which have some limitations such as small screen, limited memory, processing capacity, low-resolution displays, and data transfer speed. Therefore, most mobile-based studies highlight the positive effect of quality on users' satisfaction and users' intention to use mobile technologies [Kim *et al.*, 2010; Lee and Chung, 2009; Vlachos and Vrechopoulos, 2008; Wang and Liao, 2007]. This study focuses only on the direct influences of quality factors on intention to use MFI, even though Koivumäki *et al.* [2008] pointed out the indirect influences of quality factors on users' satisfaction toward intention to use, since they found that users' satisfaction and intention had a strong correlation ($\beta = 0.74$, $p < 0.01$). Furthermore, most previous studies reported similar influences of quality factors in both users' satisfaction and intention to use [DeLone and Mclean, 2003; Tsang *et al.*, 2011]. Kim *et al.* [2012] stated that the screen design, coordination of image, proper color, and appropriate image on the screen's device can have an impact on the intention to use mobile technology. Based on the MFI features, this study evaluated mobile design qualities of three aspects: information quality, interaction quality, and presentation quality.

Information quality is commonly acknowledged as one of the useful proxy measures of IT success [Chae *et al.*, 2002; DeLone and McLean, 1992, 2003]. Therefore, information quality should be taken into account when firms want to design and develop mobile information services for their customers. The mobile information quality of MFI was conceptualized in terms of sufficient content, updating content, completeness, accuracy, and reliability of sources [Chen, 2008; Kim and Hwang, 2012; Lee and Chung, 2009; Tsang *et al.*, 2010; Wang and Liao, 2007]. Most previous studies found that information quality positively impacted users' intention to use technologies [Kim *et al.*, 2010] and users' satisfaction [Chae *et al.*, 2002; Lee and Chung, 2009; Tsang *et al.*, 2011; Wang and Liao, 2007]. In order to explore and examine this association, the study hypothesizes the effect of mobile information quality as follows:

H5: Information quality has a positive impact on intention to use MFI.

Interaction quality has been a topic of interest in IT studies since the last decade. Interaction quality is achieved when IT provides easy and efficient methods of interaction [Chae *et al.*, 2002; Bratianu, 2015; Jennex, 2008]. In this study, the mobile interaction quality of MFI was conceptualized as its usability in terms of structure, navigation, and consistency. Structure refers to the hierarchical multilevel arrangement of menus and content in which tourists may find the desired information. Meanwhile, navigation refers to a guide (e.g., menus, buttons, titles, labels) which indicates the specific information that should guide tourists through the software without making them lose track of where they are relative to other locations. Consistency refers to consistent design of menus and buttons to help users in terms of navigation, learnability, and memorability. Chae *et al.* [2002] revealed that interaction quality influenced consumers' intention to use mobile internet services.

Likewise, Koivumäki *et al.* [2008] reported that interaction quality positively impacted users' satisfaction with mobile information services in both fun and functional environments. Thus, interaction quality possibly has a positive influence on tourists' intention to use MFI. In order to examine this association, the study hypothesizes as follows:

H6: Interaction quality has a positive impact on intention to use MFI.

Presentation quality is crucial in helping to meet users' expectations and is a significant predictor of IT's ability to fulfill users' satisfaction and of users' intention to use technologies [Kim *et al.*, 2012; Tsang *et al.*, 2010; Wang and Liao (2007)]. The presentation quality of MFI in this study was conceptualized as the quality of the displayed screen in terms of its attractiveness, organization, user-friendliness, layout, and multimedia offering. Even though Lee and Chung [2009] evidenced that interface design quality did not impact customer satisfaction with mobile banking in Korea, most prior research reported that presentation quality was associated with users' intention to use IT and mobile services. For example, Thong *et al.* [2002] showed that screen design indirectly influenced the intention to use digital libraries through perceived ease-of-use construct. Aladwani [2006] found that presentation quality indirectly influenced the intention to purchase online through the consumer's attitude towards the website, and Kim *et al.* [2012] revealed that interface design impacted usage intentions in the context of tour information services. Therefore, this study expects mobile presentation quality to possibly influence tourists' intentions to use MFI. The study proposes the following hypothesis in order to support this assumption:

H7: Presentation quality has a positive impact on the intention to use MFI.

5. Methodology

The researchers conducted a quantitative investigation using a survey questionnaire to collect data from international tourists at the Departure International Hall at Suvarnabhumi Airport, Thailand. The research methodology was designed as explained in the following.

5.1. *Measurement development*

The study employed a survey questionnaire and adapted, from previous studies, items showing Cronbach's alpha values above 0.70. The items of perceived usefulness, perceived compatibility, and perceived ease-of-use were adapted from IT researches [Brown *et al.*, 2003; Chang and Pan, 2011; Hsu *et al.*, 2007; Kim *et al.*, 2010; Lu *et al.*, 2011; Richardson, 2011]. The items of perceived mobility were adapted from mobile service research and social psychology research [Haustein and Hunecke, 2007; Kim *et al.*, 2010]. For mobile design qualities, the items of information quality and presentation quality were adapted from Wang and Liao [2007] and Tsang *et al.* [2010]. Meanwhile, the items of interaction quality were adapted from Chae *et al.* [2002]. Furthermore, items of intention to use MFI were adapted from studies of mobile services and information systems for tourists [Ayeh *et al.*, 2013; Kim, 2008; Oh *et al.*, 2009]. The questionnaire used the 5-point Likert scale ranging from 1 (strongly disagree) to 5 (strongly agree). In order to ensure that the questionnaire had been effective, a pilot test was conducted by distributing questionnaires to 30 international tourists who visited Thailand. The instrument was finalized after making minor changes based on feedback from the pilot test.

5.2. *Sample and data collection*

The population of the study was international tourists who visited Thailand. The data were collected by convenience sampling using a survey questionnaire. The respondents were international tourists in the International Departure Hall of Suvarnabhumi Airport. The total number of the final effective sample is 497, sufficient to test the hypotheses [Hair *et al.*, 2010; Sekaran and Bougie, 2010]. There were two groups of respondents; independent tourists (275 respondents) and tourists using travel agencies or package tours (222 respondents). Most of them had eaten Thai food before (89.9%) and 77 respondents had food allergies. 62.0% of respondents have experiences with mobile travel guide and 88.5% of them have experiences with mobile applications. There were 236 male respondents and 261 female respondents. Most were between 21 and 50 years old. Most respondents were Asian or European (71.2%). The detailed descriptive statistics relating to the respondents' demographic are shown in Table 2.

6. Data Analysis and Results

6.1. *Hypotheses testing*

To test the proposed hypotheses, the SEM using WarpPLS software with PLS algorithm was used to assess the relationships among the proposed constructs in the structural model. The results are summarized in Table 3. For the data of all respondents, the proposed model could explain 68.0% of the total variance exhibited in intention to use MFI. According to Chin [1998], R^2 values of 0.19, 0.33, or 0.67 for endogenous latent constructs of the structural model are described as weak, moderate, or substantial. Thus, the proposed model of study can be assumed to sufficiently reflect the international tourists' intention to use MFI. The path analysis reveals that all hypotheses were supported except H1. It implies that perceived usefulness is not a factor of intention to use MFI at the significant level 0.05. In terms of the perception factors, perceived ease-of-use, perceived compatibility, and perceived mobility had the positive impact on international tourists'

Table 2. The demographic data of respondents.

Demographics		Total		Independent tourists		Package tourists	
		Number	%	Number	%	Number	%
Gender	Male	236	47.5	130	55.1	106	44.9
	Female	261	52.5	145	55.6	116	44.4
Age (years)	< 21 years	79	15.9	56	20.4	23	10.4
	21–30 years	93	18.7	65	23.6	28	12.6
	31–40 years	110	22.1	64	23.3	46	20.7
	41–50 years	106	21.3	46	16.7	60	27.0
	51–60 years	57	11.5	21	7.6	36	16.2
	> 60 years	52	10.5	23	8.4	29	13.1

Religion	None	158	31.8	103	37.5	55	24.8
	Christianity	148	29.8	65	23.6	83	37.4
	Islam	109	21.9	62	22.5	47	21.2
	Hinduism	27	5.4	19	6.9	8	3.6
	Buddhism	16	3.2	5	1.8	11	5.0
	Chinese Folk	14	2.8	4	1.5	10	4.5
	Other	25	5.1	17	6.2	8	3.6
Continent	Asia	192	38.6	71	25.8	121	54.5
	Europe	162	32.6	104	37.8	58	26.1
	America	75	15.1	48	17.5	27	12.2
	Oceania	56	11.3	40	14.5	16	7.2
	Africa	7	1.4	7	2.5	0	0.0
	Middle East	5	1.0	5	1.9	0	0.0

Table 3. Summary of hypotheses testing results.

Hypothesis	Paths	Beta value	*t*-value	Supported
H1	PU → INU	0.17	1.08	No
H2	PEU → INU	0.30	6.75	Yes
H3	PCO → INU	0.45	6.49	Yes
H4	PMO → INU	0.28	4.17	Yes
H5	IFQ → INU	0.39	5.45	Yes
H6	IAQ → INU	0.38	2.12	Yes
H7	PSQ → INU	0.35	4.27	Yes

intention to use MFI. Perceived compatibility, especially, had the most influence on tourists' intention. In terms of mobile design quality, information quality, interaction quality, and presentation quality had positive significant effects on international tourists' intentions to use MFI.

6.2. *Comparison between independent tourists and package tourists*

The study further compared the association of constructs in the intention model between the two study groups. The associations of constructs in both groups were similar to the associations of all respondents as presented in Figure 2. However, the influence of each factor is different in each tourist group.

In Figure 2, the first value on the relationship line is the beta value of the independent tourists while the second value is the beta value of the package tourists. The proposed model was more efficient at predicting the intention to use MFI in the group of independent tourists ($Q^2 = 0.76$) than in the group of package tourist ($Q^2 = 0.58$). Among perception factors, perceived compatibility and perceived mobility had more impact on the intention to use MFI in the group of independent tourists than in the group of package tourists. In contrast, perceived ease-of-use in the group of independent tourist had less impact

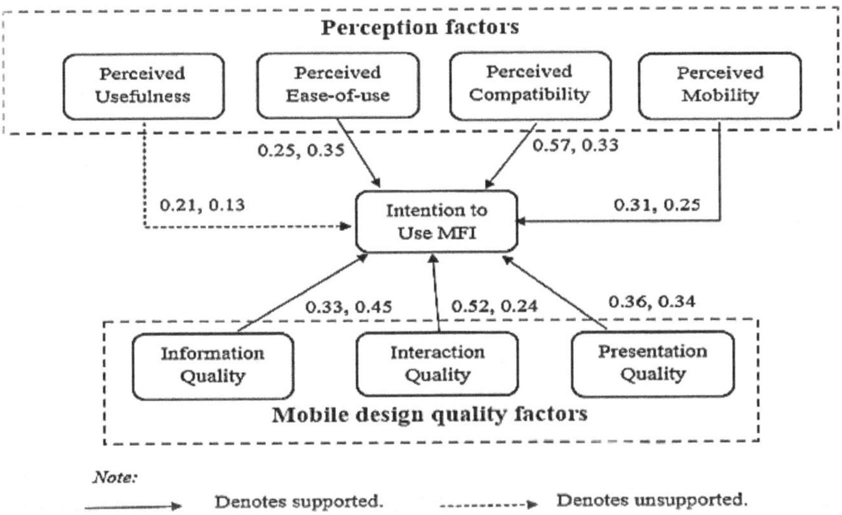

Figure 2. The comparative path analysis results.

than it had in the group of package tourists. The independent tourists were more concerned about design quality factors of interaction and presentation than package tourists. Meanwhile, the package tourists were more concerned about information quality than the other group.

To compare mean values, a *t*-test was conducted. There were significant differences in factors of international tourists' intention to use MFI between the independent tourist group and the package tourist group as presented in Table 4. For perceived compatibility, the independent tourist group showed higher mean value on "Proper information format for travel style" (t-value = 2.32, $p < 0.05$). Likewise, the "Anywhere" was the only difference in perceived mobility between both groups (t-value = 2.02, $p < 0.05$). Furthermore, the mean value of perceived ease-of-use in the independent tourist group for the "Level of effort" aspect was less than the package tourist group (t-value = –2.31, $p < 0.05$). In terms of mobile design qualities, the mean value in both groups varied significantly in the "Up-to-date" category of information quality (t-value = 2.12,

Table 4. The comparative results between independent tourist and package tourists.

Construct	Type of tour management			
	Independent tourists Mean (SD)	Package tourists Mean (SD)	Mean difference	*t*-value
Perceptions				
Perceived usefulness				
PU1: Convenience of information access	4.46 (0.89)	4.44 (0.89)	0.02	0.23
PU2: Speed of information search	**4.38 (0.86)**	**4.23 (0.95)**	**0.15**	**1.77**
PU3: Tour enhancement	**4.51 (0.93)**	**4.37 (0.89)**	**0.14**	**1.82**
Perceived compatibility				
PCO1: Sufficient information fit to travel	4.27 (0.93)	4.16 (0.89)	0.11	1.22
PCO2: Proper information format for travel	4.23 (0.93)	4.01 (1.02)	0.14	2.32*
PCO3: Recommended food fit to lifestyle	4.21 (0.92)	4.16 (0.89)	0.05	0.50
Perceived ease-of-use				
PEO1: Ease of learning	4.60 (0.76)	4.47 (0.66)	0.13	1.78
PEO2: Ease of usage	4.65 (0.58)	4.65 (0.52)	0.00	0.12
PEO3: Level of effort	4.48 (0.68)	4.62 (0.70)	–0.14	–2.31*
Perceived mobility				
PMO1: Anytime	4.38 (0.80)	4.23 (0.80)	0.15	1.89
PMO2: Anywhere	4.38 (0.78)	4.22 (0.80)	0.16	2.02*
PMO3: Suitable tour information on movement	4.46 (0.80)	4.44 (0.71)	0.02	0.22

Mobile design qualities				
Information quality				
IFQ1: Sufficient	4.11 (0.94)	3.97 (0.97)	0.14	1.49
IFQ2: Up-to-date	4.22 (0.87)	4.04 (0.89)	0.18	2.12*
IFQ3: Complete	3.95 (1.06)	3.92 (1.00)	0.03	0.30
IFQ4: Accurate	4.05 (1.03)	4.04 (0.89)	0.01	0.06
IFQ5: Good source of information	4.01 (1.03)	4.08 (0.93)	–0.07	–0.70
Interaction quality				
IAQ1: Menu categories	4.57 (0.70)	4.31 (0.80)	0.26	3.51***
IAQ2: Location recognition	4.48 (0.75)	4.45 (0.69)	0.03	0.35
IAQ3: Go back previous page	4.51 (0.74)	4.37 (0.78)	0.14	1.79
IAQ4: Menu awareness	4.48 (0.74)	4.34 (0.74)	0.14	1.87
IAQ5: Consistent contents	4.35 (0.78)	4.36 (0.73)	–0.01	–0.16
Presentation quality				
PSQ1: Attractiveness	4.42 (0.79)	4.40 (0.73)	0.02	0.15
PSQ2: Organized information	4.57 (0.63)	4.52 (0.61)	0.05	0.91
PSQ3: User-friendly	4.59 (0.61)	4.63 (0.53)	–0.04	–0.67
PSQ4: Appropriate layout	4.53 (0.69)	4.53 (0.59)	0.00	–0.00
PSQ5: Proper multimedia features	4.47 (0.75)	4.34 (0.88)	0.13	1.59
Intention to use MFI				
INU1: Before trip: Trip planning	4.39 (0.81)	4.34 (0.83)	0.05	0.64
INU2: During trip: Information usage	4.54 (0.75)	4.50 (0.63)	0.04	0.55
INU3: After trip: Sharing trip experiences	4.13 (1.10)	4.09 (1.05)	0.04	0.42
INU4: Next trip	4.53 (0.73)	4.52 (0.63)	0.01	0.29

$p < 0.05$) and "Menu categories" of interaction quality (t-value = 3.51, $p < 0.001$).

7. Discussions and Findings

This study introduced additional factors into the TAM by using PLS estimation to investigate tourists' intention to use MFI for travel. Comparing the results with previous TAM researches reveals several interesting findings. In the context of MFI, there was no relationship between perceived usefulness and the intention to use MFI. This implies that perceived usefulness is not a factor of tourists' intention to use MFI. This finding is contrary to most prior studies in other contexts [Ayeh *et al.*, 2013; Chang and Pan, 2011; Gerpott, 2011; Kim and Shin, 2015; Yadav *et al.*, 2016] and TAM. The possible causes are consumed behavior, belief, and religion. International tourists will not use any useful information which does not support their consumed behaviors, beliefs, and religions. However, the study found significant support for a relationship between perceived ease-of-use and intention to use MFI. This result is consistent with the TAM and most previous studies [Brown *et al.*, 2003; Kim and Shin, 2015; Kim *et al.*, 2010; Xue *et al.*, 2012]. The finding also supports the report of Rukzio *et al.* [2008] that the usage of mobile devices requires additional cognitive and mental efforts because of the limited input capability and small screen. Therefore, perceived ease-of-use has a positive impact on international tourists' intention to use MFI. Similarly, perceived compatibility has a significant effect on international tourists' intention to use in the context of MFI. This finding corresponds to prior researches [Gerpott, 2011; Schierz *et al.*, 2010; Xue *et al.*, 2012] which pointed out the important role of perceived compatibility on intention to use technology. In terms of perceived mobility, this study confirms previous studies that perceived mobility to be the key to mobile service adoption and an influence on the intention to use technology [Kaba and Osei-Bryson, 2009; Schierz *et al.*, 2010]. As the results of travel behavior, tourists expect to access information and services through their mobile devices anytime and anywhere during their travel aboard. They want recommendations of food and restaurants based on their existing behaviors. This finding confirms the statement of Kim *et al.* [2010] that mobile users will expect benefits in terms of information access, communication, and service anytime and anywhere.

Considering quality factors, there are three qualities affecting tourists' intention to use MFI. The first one is information quality which was identified as one of the useful proxy measures of IT success by Chae *et al.* [2002] and DeLone and Mclean [2003]. The result depicts the positive impact of information quality on intention to use MFI. This finding is consistent with the studies of Kim *et al.* [2010] and Lee and Chung [2009]. It implies that international tourists will use MFI when the MFI has adequate information in terms of content, updating content, completeness, accuracy, and reliability [Chen, 2008; Kim and Hwang, 2012; Lee and Chung, 2009; Tsang *et al.*, 2010; Wang and Liao, 2007]. The second quality is interaction which is measured by navigation, structural level, and design consistency. In the context of MFI, the finding reveals the positive relationship between the interaction quality and tourists' intention to use. International tourists are willing to use MFI when the interaction quality they receive is convenient and comfortable. This finding is similar to the study in other context of Chae *et al.* [2002] that highlighted the positive influence of interaction quality on customers' intention to use. The last quality is presentation which is identified as being crucial in helping to meet users' expectations and a significant predictor of IT, by Tsang *et al.* [2010] and Wang and Liao [2007]. This relationship between presentation quality and intention to use MFI was also found in this study. In response to the screen size, international tourists expect good quality in terms of attractiveness, organization, user-friendliness, screen layout, and proper multimedia. This finding supports the study of Kim *et al.* [2012] that revealed the direct impact of interface design on usage intention in the context of tour information services.

8. Contributions, Limitations, and Future Research

This study proposed a research model based on the TAM [Davis, 1989]. The proposed model was empirically tested for its capacity to explain an individual's intention to use mobile IT applications in the context of MFI. There are some research contributions of interest for researchers, practitioners, and tourism providers. This study was able to advance the literature on technology adoption and facilitate tourism providers in MFI.

8.1. *Theoretical implications*

From the theoretical perspective, this study has contributed to the existing literature by identifying the key factors affecting tourists' intention to use MFI. Taking into consideration the theoretical backgrounds of TAM and mobile design qualities, this study proposed a research model to provide an understanding of the motives for adoption of MFI among international tourists. The research revealed that intention to adopt MFI is affected by perception factors (i.e., perceived ease-of-use, perceived compatibility, and perceived mobility) and mobile design qualities (i.e., information quality, interaction quality, and presentation quality). Furthermore, the results indicate the most important influence on international tourists' intention to use MFI is perceived compatibility. However, there is a departure from the expected results associated with perceived usefulness and tourists' intention to use MFI. International tourists not focus only on the usefulness of food information, but they focus on the food information and appropriate mobile-based technology which supports their travel behaviors and consumed behaviors. As presented in Table 4, international tourists prefer to use MFI during the course of their travel rather than before or after. It implies that international tourists usually use MFI when traveling abroad because of their different food cultures and requirements from recommended restaurants. Therefore, future mobile-based research should not ignore perceived compatibility as a factor of international tourists' intention to use food information.

In terms of measurement findings, the results of this study suggest that the measurement of intention to use mobile IT in the context of tourism should evaluate the users' intention in three phases of travel: before the trip, during the trip, and after the trip. In the measurement analysis of this study, the item loading values of intention to use MFI are acceptable and reliability values are good. The results of this study extend the traditional measurement of intention to use technology which evaluates only "Intention to use it in the future" and "Intention to use it the next time" [David, 1989; Rogers, 2003]. Therefore, future research into tourists' intention to use technology should evaluate the tourists' intention to use IT in three travel phases (before the trip, during the trip, and after the trip) including intention to use IT in the next time.

With regards the comparative findings between the two types of tourist, the influences of each factor on tourists' intention to use MFI are similar. Perceived usefulness does not influence tourists' intention to use MFI, and perceived compatibility has the most influence on tourists' intention to use MFI as presented in Figure 2. Furthermore, the findings reveal that the proposed model can explain and predict international tourists' intention to use MFI among groups of independent tourists and package tourists. In addition, the findings also demonstrate some differences between the two types of tourist in terms of perceptions (i.e., proper information format for travel style, level of effort, and availability anywhere) and mobile design quality (i.e., updating information and proper menu categories). Therefore, these findings provide a useful guideline to help researchers who want to study these two types of tourists with regard to their intention to use food IT.

8.2. *Practical implications*

The results suggest that the strongest predictor of international tourists' intention to use MFI is perceived compatibility. This implies the provision of good food information which is sufficiently well-formatted to support tourists' travel style, and food and restaurant recommendations that respond to their lifestyles. Based on that, this study suggested that tourism providers and MFI developers should take into account the following suggestions when providing food information for international tourists: (1) It is undesirable to provide a lot of food information for them because mobile devices have small screens and reading a lot of text can be difficult; (2) The food information provided should be tailored to suit international tourists' requirements during their time abroad more than before travel and after traveling, because most use MFI to inform themselves about local food ingredients, recommended menus, and restaurants that support their food allergies and religious demands, and nearby restaurants; (3) The MFI should present good categorized screen menus, proper multimedia formats, proper and consistent screen design. These will help tourists to easily use and interact with the MFI.

The findings can help tourism providers and MFI developers differentiate between the needs of independent tourists and package tourists, saving money and time in the provision and implementation of MFI. If the MFI is intended for use by both tourist groups, they can ensure that the factors of international tourists' intention to use MFI are the same in terms of perception and mobile design quality. However, if targeting only independent tourists or only package tourists, then they should provide food information which is focused on their different needs as presented in Table 4. For independent tourists, the MFI should provide up-to-date recommended menus based on their consumption requirements (i.e., food allergy, belief, and religion) and restaurants near their current location. With regard to their behavior, they usually use the MFI by themselves to learn about local food during their travel abroad. Therefore, the food information should be displayed in menu categories that can support interaction on mobile devices. Based on tourists' suggestions in the questionnaire, the MFI should be categorized according to the cooking method (i.e., boiling, stirring, frying, steaming, and roasting) or flavor (i.e., salty, sweet, spicy, sour, and mild). In addition, the page of food information should have a link button to Google Maps in order to receive directions and the route to their selected restaurant. However, they cannot connect to the Internet everywhere in Thailand, thus the MFI should be available in offline mode as well as online. In particular, a graphic map and text directions to their selected restaurant should be provided wherever GPS is unavailable. For the package tourist groups, most of their meals during travel abroad are supported by their tourist agencies and most of them have less experience of using mobile tourist guides than the independent international tourists. They use MFI only sometimes or when they have free time from their tour schedule. Therefore, they expect an application which is easy to use on their mobile device. Tourism providers and MFI developers should consider ease-of-use and learning in their implementation process.

8.3. *Limitation and future research*

Future research could go beyond the concept of international tourists' intention to use MFI for travel aboard and consider the factors (i.e., perceptions and mobile design qualities) in the model that influences tourists' intention to use MFI. However, this study measures the tourist's intention to use MFI based on their travel behaviors today. In the future, if they have more occasion to use MFI in three travel phases or more, then future research should measure more tourists' travel behavior in that time. The example is that most international tourists enjoy sharing their experiences during and after a trip via social media.

If mobile technology changes in the future, then tourists' activities after trips may not only be about sharing experiences. Furthermore, the study identifies the key factors associated with the intention to use MFI among international tourists in Thailand. The percentages of respondents in this study are followed by international tourist arrivals to Thailand by continent. Therefore, future research should empirically verify the validity of the proposed model across different groups of respondents; in particular, international tourists from the Middle East and/or Africa.

Acknowledgments

The work was supported through government funding by Prince of Songkla University (SCI580883S) and INTACT Research Unit. The authors wish to express their gratitude to Suvarnabhumi Airport for permission to collect data within the airport's precincts.

References

Agrebi, S. and Jallais, J. (2015). Explain the intention to use smartphones for mobile shopping. *Journal of Retailing and Consumer Services*, **22**:16–23.

Aladwani, A. M. (2006). An empirical test of the link between website quality and forward enterprise integration with web consumers. *Business Process Management Journal*, **12**(2):178–190.

Ayeh, J. K., Au, N. and Law, R. (2013). Predicting the intention to use consumer-generated media for travel planning. *Tourism Management*, **35**:132–143.

Bratianu, C. (2015). *Organizational Knowledge Dynamics: Managing Knowledge Creation, Acquisition,Sharing, and Transformation*. IGI Global, NY.

Brown, I., Cajee, Z., Davies, D. and Stroebel, S. (2003). Cell phone banking: Predictors of adoption in South Africa-An exploratory study. *International Journal of Information Management*, **23**(5): 381–394.

Chae, M., Kim, J., Kim, H. and Ryu, H. (2002). Information quality for mobile Internet services: A theoretical model with empirical validation. *Electronic Markets*, **12**(1):38–46.

Chang, S. E. and Pan, Y. H. V. (2011). Exploring factors influencing mobile users' intention to adopt multimedia messaging service. *Behaviour and Information Technology*, **30**(5):659–672.

Chen, L. (2008). A model of consumer acceptance of mobile payment. *International Journal of Mobile Communications*, **6**(1):32–52.

Chin, W. W. (1998). *The Partial Least Squares Approach for Structural Equation Modeling*, Lawrence Erlbaum Associates, United States.

Chung, J. Y. and Buhalis, D. (2008). Web 2.0: A study of online travel community. *Information and Communication Technologies in Tourism*. Springer, Innsbruck, Austria, 2008, pp. 70–81.

Dai, H. and Palvi, P. C. (2009). Mobile commerce adoption in China and the United States: A cross-cultural study. *ACM SIGMIS Database*, **40**(4):43–61.

Davis, F. (1989). Perceived usefulness, perceived ease of use, and user acceptance of information technology. *MIS Quarterly*, **13**(3): 319–340.

DeLone, W. H. and McLean, E. R. (1992). Information systems success: The quest for the dependent variable. *Information Systems Research*, **3**(1):60–95.

DeLone, W. H. and McLean, E. R. (2003). The DeLone and McLean model of information systems success: A ten-year update. *Journal of Management Information Systems*, **19**(4):9–30.

Feng, S., Feng, Z. and Jin, R. (2015). Learning to rank image tags with limited training examples. *IEEE Transactions on Image Processing*, **24**(4):1223–1234.

Flurry Insights (2015). Mobile use grows 76% year-over-year. Available at http://flurrymobile.tumblr.com/post/115194992530/shopping-productivity-and-messaging-give-mobile.

Gerpott, T. J. (2011). Attribute perceptions as factors explaining mobile internet acceptance of cellular customers in Germany — An empirical study comparing actual and potential adopters with distinct categories of access appliances. *Expert Systems with Applications*, **38**(3):2148–2162.

Hair, J. F., Black, W. C., Babin, B. J. and Anderson, R. E. (2010). *Multivariate Data Analysis*, 7th edn. Pearson Prentice Hall, New Jersey.

Haustein, S. and Hunecke, M. (2007). Reduced use of environmentally friendly modes of transportation caused by perceived mobility necessities: An extension of the theory of planned behavioral. *Journal of Applied Social Psychology*, **37**(8):1856–1883.

Hill, R. and Troshani, I. (2007). A conceptual framework and propositions for the acceptance of mobile services. *Journal of Theoretical and Applied Electronic Commerce Research*, **2**(2):61–73.

Hsu, C. L., Lu, H. and Hsu, H. H. (2007). Adoption of the mobile Internet: An empirical study of multimedia message service (MMS). *Omega*, **35**(6):715–726.

Jennex, M. E. (2008). *Current Issues in Knowledge Management*. IGI Global, NY.

Jung, J., Chan-Olmsted, S., Park, B. and Kim, Y. (2012). Factors affecting e-book reader awareness, interest, and intention to use. *New Media & Society*, **14**(2):204–224.

Kaba, B. and Osei-Bryson, K. M. (2009). External factors influencing mobile phones use in Quebec: An exploratory study. *The 15th Americas Conference on Information Systems, AMCIS 2009*, AIS Electronic Library, pp. 1–14.

Kim, C., Mirusmonov, M. and Lee, I. (2010). An empirical examination of factors influencing the intention to use mobile payment. *Computers in Human Behavior*, **26**(3):310–322.

Kim, D. J. and Hwang, Y. (2012). A study of mobile Internet user's service quality perceptions from a user's utilitarian and hedonic value tendency perspectives. *Information Systems Frontiers*, **14**:409–421.

Kim, D. Y., Park, J. and Morrison, A. M. (2008). A model of traveller acceptance of mobile technology. *International Journal of Tourism Research*, **10**(5):393–407.

Kim, J., Ahn, K. and Chung, N. (2012). Examining the factors affecting perceived enjoyment and usage intention of ubiquitous tour information services: A service quality perspective. *Asia Pacific Journal of Tourism Research*, 1–20.

Kim, K. J. and Shin, D. H. (2015). An acceptance model for smart watches: Implications for the adoption of future wearable technology. *Internet Research*, **25**(4):527–541.

Kim, S. H. (2008). Moderating effects of job relevance and experience on mobile wireless technology acceptance: Adoption of a smartphone by individuals. *Information and Management*, **45**(6): 387–393.

Koivumäki, T., Ristola, A. and Kesti, M. (2008). The effects of information quality of mobile information services on user satisfaction and service acceptance: Empirical evidence from Finland. *Behavior and Information Technology*, **27**(5):375–385.

Kuo, R. Z. and Lee, G.-G. (2011). Knowledge management system adoption: Exploring the effects of empowering leadership, task-technology fit and compatibility. *Behaviour & Information Technology*, **30**(1):113–129.

Lawson-Body, A., Willoughby, L., Mukankusi, L. and Logossah, K. (2012). Adoption of cell phones as preparedness efforts for natural disasters. *Issues in Information Systems*, **13**(1):11–20.

Lee, K. C. and Chung, N. (2009). Understanding factors affecting trust in and satisfaction with mobile banking in Korea: A modified DeLone and McLean's model perspective. *Interacting with Computers*, **21**(5):385–392.

Lu, Y., Yang, S., Chau, P. Y. K. and Cao, Y. (2011). Dynamics between the trust transfer process and intention to use mobile payment services: A cross-environment perspective. *Information and Management*, **48**:393–403.

Luo, X., Gurung, A. and Shim, J. (2010). Understanding the determinants of user acceptance of enterprise instant messaging: An empirical study. *Journal of Organizational Computing and Electronic Commerce*, **20**(2):155–181.

Oh, S., Lehto, X. Y. and Park, J. (2009). Travelers' intent to use mobile technologies as a function of effort and performance expectancy. *Journal of Hospitality Marketing & Management*, **18**(8): 765–781.

Peres, R., Correia, A. and Moital, M. (2011). The indicators of intention to adopt mobile electronic tourist guides. *Journal of Hospitality and Tourism Technology*, **2**(2):120–138.

Pham, T. T. T. and Ho, J. C. (2015). The effects of product-related, personal-related factors and attractiveness of alternatives on consumer adoption of NFC-based mobile payments. *Technology in Society*, 43:159–172.

Phan, K. and Daim, T. (2011). Exploring technology acceptance for mobile services. *Journal of Industrial Engineering and Management*, **4**(2):339–360.

Richardson, J. W. (2011). Technology adoption in Cambodia: Measuring factors impacting adoption rates. *Journal of International Development*, **23**(5):697–710.

Rogers, E. M. (2003). *Diffusion of Innovations*, 5th edn. Free Press, NY.

Rukzio, E., Noda, C., De Luca, A., Hamard, J. and Coskun, F. (2008). Automatic form filling on mobile devices. *Pervasive and Mobile Computing*, **4**(2):161–181.

Schierz, P., Schilke, O. and Wirtz, B. (2010). Understanding consumer acceptance of mobile payment services: An empirical analysis. *Electronic Commerce Research and Applications*, **9**(3):209–216.

Sekaran, U. and Bougie, R. (2010). *Research Methods for Business: A Skill Building Approach*, 5th edn. John Wiley and Sons, United Kingdom.

Simon-Wong, C. K. and Gladys-Liu, F. C. (2011). A study of pre-trip use of travel guidebooks by leisure travelers. *Tourism Management*, **32**(3):616–628.

Tan, M. and Teo, T. S. H. (2000). Factors influencing the adoption of Internet banking. *Journal of the AIS*, **1**(5):1–44.

Thong, J. Y. L., Hong, W. and Tam, K. Y. (2002). Understanding user acceptance of digital libraries: What are the roles of interface characteristics, organizational context, and individual differences? *International Journal of Human–Computer Studies*, **57**(3): 215–242.

Tsang, N. K. F., Chan, G. K. Y. and Ho, K. K. F. (2011). A holistic approach to understanding the use of travel guidebooks: Pre-, during, and post-trip behavior. *Journal of Travel & Tourism Marketing*, **28**(7): 720–735.

Tsang, N. K. F., Lai, M. T. H. and Law, R. (2010). Measuring e-service quality for online travel agencies. *Journal of Travel and Tourism Marketing*, **27**(3):306–323.

Vlachos, P. A. and Vrechopoulos, A. P. (2008). Determinants of behavioral intentions in the mobile internet services market. *Journal of Services Marketing*, **22**(4):280–291.

Wang, Y. S. and Liao, Y. W. (2007). The conceptualization and measurement of m-commerce user satisfaction. *Computers in Human Behavior*, **23**(1):381–398.

Xue, L., Yen, C. C., Chang, L., Chan, H. C., Tai, B. C., Tan, S. B. and Choolani, M. (2012). An exploratory study of ageing women's perception on access to health informatics via a mobile phone-based intervention. *International Journal of Medical Informatics*, **81**:637–648.

Yadav, R., Sharma, S. K., Tarhini, A. and Irani, Z. (2016). A multi-analytical approach to understand and predict the mobile commerce adoption. *Journal of Enterprise Information Management*, **29**(2): 222–237.

Yu, C. C. and Chang, H. P. (2009). Personalized location-based recommendation services for tour planning in mobile tourism applications. E-Commerce and Web Technologies, (EC-Web 2009) Springer, Linz, Austria, 2009, pp. 38–49.

Chapter 9

Emerging Applications: Tourism

Numtip Trakulmaykee and Parichard Benrit

1. Introduction

Mobile information technology (IT) is currently growing rapidly in a variety of contexts such as m-commerce, m-banking, m-learning, m-health, m-guide, and m-book. With the use of application in mobile devices, it becomes easy to gain information anywhere and anytime. As a result of features and utilities of smartphones, mobile applications have become popular personal information processors, today. The mobile information application also becomes a technological trend that many businesses adopt to give information and advertisement directly to their customers. In addition, the high growth of mobile phone subscribers worldwide also increases the use of mobile applications, as presented in Figure 1 [Flurry Analytic, 2012]. Comparing 1st quarter of 2011 with 1st quarter of 2012, the graph shows the 10 fastest growing countries in terms of the launching and the use of mobile application. China leads the world in mobile application growth, with an enormous growth rate of more than 1,000% in this period. Furthermore, the figures also report the high growth rate of several countries in Asia such as Philippines, India, and Thailand. Therefore, nowadays, many businesses are adopting mobile applications in their marketing to gain benefit, especially the businesses which serve international customers.

However, the mobile IT adoption based on the IT features, or innovation characteristics, and target users is not easy to

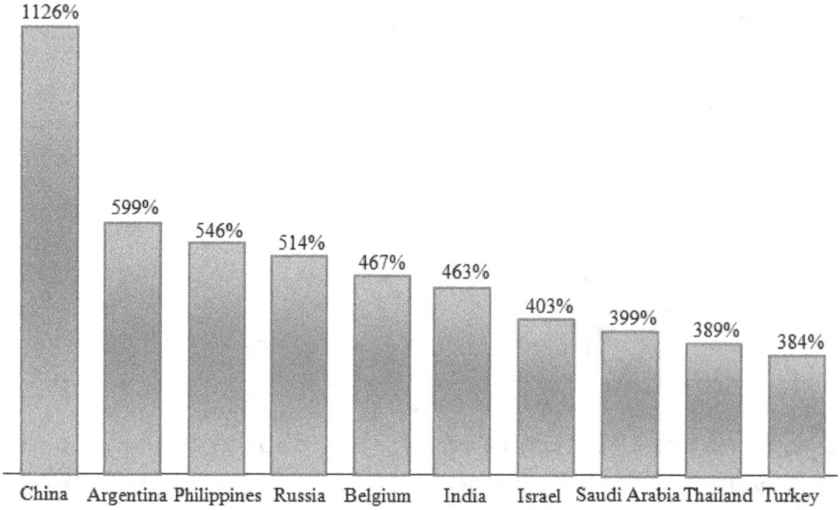

Figure 1. Comparison of mobile application growth by country in 1st quarter between 2011 and 2012.

achieve. In order to achieve the successful mobile IT adoption, the businesses and previous studies are concerned about users' intention to use IT and the determinants of users' intention to use IT [Kim and Kizildag, 2011; Schierz *et al.*, 2010; Wu *et al.*, 2011]. From previous IT studies, we found a few research gaps. First, the literature on modern mobile-based mostly looks at the innovation characteristics of IT as factors of users' intention to use IT in the last decade [Brown *et al.*, 2003; Mallat, 2007; Kim and Kizildag, 2011]. Second, the relationship between innovation characteristics and users' intention to use IT is inconsistent in the different technologies and user groups. For example, Dai and Palvi [2009] found the impact of compatibility on users' intention to use mobile commerce in the United States, while it did not impact on users' intention mobile commerce in China, whereas Putzer and Park [2010] found that compatibility was a factor of user intention to use Health Information Technology (HIT) on smartphones in the United States. Finally, several previous mobile IT studies have emphasized the interaction quality as the determinants of users' satisfactions or users' intention [Chae *et al.*, 2002; Koivumaki *et al.*, 2008; Lin *et al.*, 2012], due to the mobile device limitations such as small screen, limited memory, processing capacity, low-resolution displays, and data transfer speed. However, the interaction quality has still be not been studied in terms of moderating variable, which possibly strengthens or weakens the relationship between innovation characteristics and users' intention to use mobile IT.

In tourism industry, the mobile IT also has been developed for travel and tourism sectors. For example, mobile hotel booking, mobile travel magazine, mobile flight ticket, mobile tour services, and mobile tourism guide (MTG). As a result of a few studies in the context of intention to use MTG among international tourists, the international tourists' intention to use MTG was considered in this study. To find out the innovation characteristics in the context of MTG, this study investigated the innovation characteristics as determinants of international tourists' intention to use MTG. In addition, the study also examined the roles of interaction quality as moderator on the relationships between innovation characteristics (relative advantage, compatibility, lack of complexity, and trialability) and tourists' intention to use MTG. The MTG of this study refers to a mobile application for international tourists to get the tour information. For instance, attractive places, maps, dos and do-not-dos during travel, climate and seasons, culture, transportation, accommodation, local food and fruits, easy native dialogues for international tourists, and currency exchange rate. The findings of this study are clear to understand and can be the evidence in terms of the innovation characteristics as determinants of international tourists' intention to use MTG. Furthermore, this study extends the body of knowledge of users' intention in the mobile IT research by including interaction quality as moderating variable.

2. Literature Review

Most previous adoption studies which investigated the determinant of users' intention to use IT are based on the three famous theories and model: the technology acceptance model (TAM), the unified theory of acceptance and use of technology (UTAUT), and the innovation diffusion of theory (IDT).

The TAM [Davis, 1989] probably is the most popular theory explaining user acceptance and behavior intention related to new technology. It provided two independent variables which measure the direct and indirect impacts on user intention to use technology such as the perceived usefulness and perceived ease-of-use. The perceived usefulness was defined as the degree to which a person believes that using a particular system would enhance his or her job performance. Meanwhile, the perceived ease-of-use was defined as the degree to which a person believes that using a particular system would be free of effort.

The UTAUT [Venkatesh *et al.*, 2003] is one of the developments in the field of general TAMs. The theory holds that four constructs play the significant role as direct determinants of user acceptance and usage behavior. Those core four constructs are performance expectancy, effort expectancy, social influence, and facilitating conditions. The performance expectancy was defined as the degree to which an individual believes that using the system will help him or her to attain gains in job performance. The effort expectancy was defined as the degree of ease associated with the use the system. The social influence was defined as the degree to which an individual perceives that important others believe he or she should use the new system. The facilitating condition was defined as the degree to which an individual believes that an organizational and technical infrastructure exists to support use of the system.

The IDT [Rogers, 2003] has been used since the 1960s to explain the process of innovation adoption. It proposed an array of innovation characteristics that may impact a users' perception of the innovation preceding the adoption. Following an extensive review of the literature, Rogers [2003] found five innovation characteristics that consistently proved to be determinants of the technological diffusion such as relative advantage, compatibility, complexity, observability, and trialability. The measurement of these characteristics was developed by Moore and Benbasat [1991], which has been widely employed in the IDT researches.

From review of these three theories and model, we found two interesting issues. First, the two constructs in these three theories and model are similar. The first construct is the perceived usefulness in TAM which is similar to the relative advantage in IDT and the performance expectancy in UTAUT. The second construct is the perceived ease of use in TAM which is obviously close to the complexity construct in IDT and the effort expectancy in UTAUT. Thus, those two constructs probably influence the users' intention to use technology in several mobile IT contexts [Kim *et al.*, 2010; Oh *et al.*, 2009; Xue *et al.*, 2012]. Second, several mobile IT researches also included the compatibility and trialability in their studies as the factors of users' intention. For instance, Kim and Kizildag [2011] integrated the constructs from TAM, IDT (compatibility and trialability), and TPB to their research model in the context of m-learning, and Putzer and Park [2010] proposed their research model based on TAM and IDT (compatibility) in the context of HIT on smartphones, and Schierz *et al.* [2010] studied users' intention to use mobile payment based on TAM and included perceived compatibility in their conceptual model.

In order to clarify the innovation characteristics as the determinants of users' intention in the context of MTG, this study investigated the determinants of tourists' intention to use MTG based on IDT. However, the study did not include the observability in the research model due to two reasons. First,

Rogers [2003] indicated that software dominant innovations have less observability than hardware, thus it appears that the more potential adopter can see an innovation. Second, Moore and Benbasat [1991] found that the observability had construct ambiguity problem in their examination. They stated that the original construct was quite complex. Hence, in this study, we did not include observability in the research model. In addition, we combined interaction quality to our model as the moderating variable. The definitions of constructs in this study were defined based on Rogers [2003], Davis [1989], and Akter *et al.* [2010] as presented in Table 1.

Table 1. The definition of the constructs in the research model.

Construct	Definition
Relative advantage	Degree to which an innovation is perceived as better than available media or system, expressed in the convenience, reducing the limitation of time and place, and enhancing effectiveness.
Compatibility	Degree to which an innovation is perceived as being consistent with the existing values, needs, and past experiences.
Lack of complexity	Degree of ease, which a tourist perceives when he or she uses the MTG in terms of learning and while using it.
Trialability	Degree to which individuals has access to the MTG, before the actual implementation in terms of illustration of using it and the chance of using trial demo version.
Interaction quality	The achievement of mobile IT providing easy and efficient methods of interaction.
Intention to use MTG	The extent to which prospective tourists intend to use MTG in three phases of traveling such as before travel, during travel, and after travel.

3. Hypotheses

The eight hypotheses were proposed to study the relationships between four innovation characteristics of MTG (relative advantage, compatibility, lack of complexity, trialability) and intention to use MTG. Additionally, the moderating effects of interaction quality were examined on those relationships.

3.1. *Relative advantage and intention*

As mention in literature review, the relative advantage construct in IDT is similar to perceived usefulness construct in TAM and performance expectancy construct in UTAUT. Hence, the relationship reviews of this study point to these constructs and intention to use IT. Most empirical studies which investigated the relationship between relative advantage and intention to use IT, found a positive influence [Ayeh *et al.*, 2013; Kim *et al.*, 2010; Xue *et al.*, 2012]. In the other word, higher relative advantage will increase the willingness to use IT. However, Hsu *et al.* [2007] found relative advantage did not influence on intention to use MMS in laggards group; while the other group (early-adopter, early-majority, and lately-majority) found the contrary results.

Therefore, this study purposed the hypothesis in order to clearly understand the relationship between relative advantage and international tourists' intention to use MTG. The hypothesis was proposed based on most previous studies results and the three famous theories and model (TAM, IDT, and UTAUT) which found relative advantage positively impacted on users' intention as the following:

H1: Relative advantage has a positive impact on intention to use MTG.

3.2. *Compatibility and intention*

Commonly, higher level of compatibility is associated with increasing intention to use IT [Baird *et al.*, 2011; Jung *et al.*,

2012; Lu *et al.*, 2011]. In the context of mobile IT and mobile services, users' lifestyle also strongly affect their decision to use the mobile technology [Baird *et al.*, 2011; Lu *et al.*, 2011; Putzer and Park, 2010; Xue *et al.*, 2012]. Because mobile IT is an extension of existing information media and services, people who frequently use mobile applications or mobile services may have less resistance to accepting the mobile version. However, some previous studies found contrary result that the compatibility did not influence the intention to use IT [Brown *et al.*, 2003; Dai and Palvi, 2009; Hsu *et al.*, 2007; Mallat, 2007].

In this study, compatibility was proposed to have a positive impact on international tourists' intention to use MTG, even though some prior studies found compatibility did not influence the intention to use IT. Most previous studies also found the relationship between compatibility and users' intention. Furthermore, Rogers [2003] stated that compatibility has a relative importance in predicting rate of adoption. Thus, in order to confirm and clearly understand the relationship between compatibility and intention to use MTG among international tourists, the hypothesis is formulated as the following:

H2: Compatibility has a positive impact on intention to use MTG.

3.3. *Lack of complexity and intention*

Many studies over the past decade have pointed to evidence regarding the critical effect of the lack of complexity on users' intention. As mentioned in literature review, the lack of complexity construct is defined similarly as perceived ease-of-use construct in TAM, effort expectancy construct of UTAUT, and complexity construct of IDT.

According to IDT of Rogers [2003], the complexity has a negative impact on rate of adoption. Therefore, the higher levels of lack of complexity are significantly associated with increased intention to use IT. Furthermore, most prior researches pointed out lack of complexity as an important factor influencing users' intention to adopt mobile IT [Kim *et al.*, 2010; Oh *et al.*, 2009; Xue *et al.*, 2012]. However, several previous studies found contrary results in this relationship [Brown *et al.*, 2003; Hsu *et al.*, 2007; Liao and Lu, 2008; Mallat, 2007].

Even though some previous studies did not find the relationship between lack of complexity and users' intention to use IT, this study found that most previous studies reported this relationship in their works. Based on those arguments and IDT [Rogers, 2003], the following hypothesis is postulated:

H3: Lack of complexity has a positive impact on intention to use MTG.

3.4. *Trialability and intention*

Rogers [2003] stated that the trialability is positively correlated with the rate of adoption. In the other word, trialability is about how easily an innovation may be experimented with as it is being adopted. If a user has a hard time in using and trying an innovation, the individual will be less likely to adopt it [Rogers, 2003].

Several previous studies found a positive impact of trialability on users' intention to use IT [Askarany *et al.*, 2007; Jung *et al.*, 2012; Ndubisi and Sinti, 2006], and users' intention to use mobile IT [Brown *et al.*, 2003; De Marez *et al.*, 2007]. While, Hernandez and Mazzon [2007] found that trialability had a negative impact on users' intention to use internet banking in Brazil. Likewise, Gerpott [2011] also found a negative impact of trialability on users' intention to use mobile internet in German. However, some IT studies reported that trialability did not have the impact on users' intention to use IT [Baird *et al.*, 2011; Hsu *et al.*, 2007; Liao and Lu, 2008]. In addition, this study also found that some previous studies which studied mobile IT and mobile service did not include trialability construct in their studies [Carter and Belanger, 2004; Mallat, 2007; Xue *et al.*, 2012].

Hence, in order to clarify the relationship between trialability and tourists' intention and impact's direction of trialability on users' intention to use in the context of MTG, the hypothesis is proposed as the following:

H4: Trialability has a positive impact on intention to use MTG.

3.5. *Moderating roles of interaction quality*

Interaction quality can be measured by three dimensions: structure, navigation, and presentation [Chae *et al.*, 2002; Koivumaki *et al.*, 2008; Park *et al.*, 2008]. Structure refers to self-descriptive indication where the specific information is located, such that users may easily find and access it. Navigation refers to interface elements that guide the users through the information space without making them lose track of where they are relative to other locations. Presentation refers that the screen is understandable, and information is provided by consistent content.

Most previous researchers studied interaction quality as an independent variable or dimension, and found it is significantly related to users' satisfaction and users' intention [Akter *et al.*, 2010; Chae *et al.*, 2002; Koivumaki *et al.*, 2008; Kuo *et al.*, 2009]. Additionally, prior studies also found interaction quality had the impact on innovation characteristics. For example, Thong *et al.* [2002] found navigator and screen design (interaction quality) influenced the perceived ease-of-use (lack of complexity) in digital library context, and Tan and Chou [2008] found interaction quality impacted on technology compatibility in the context of mobile information and entertainment service. In this study, interaction quality is identified as a moderator because when it is high, the innovation characteristics possibly influence more on tourist intention.

Hence, it is imperative to examine the roles of interaction quality on relationship between innovation characteristics and intention to use MTG, thus we hypothesized the following:

H5: The relationship between relative advantage and intention to use MTG will be stronger when interaction quality is higher.
H6: The relationship between compatibility and intention to use MTG will be stronger when interaction quality is higher.
H7: The relationship between lack of complexity and intention to use MTG will be stronger when interaction quality is higher.
H8: The relationship between trialability and intention to use MTG will be stronger when interaction quality is higher.

4. Methodology

4.1. *Population and sample*

The population of this study is the international tourists who visited Thailand. According to Sekaran and Bougie [2010], for the population above 1 million individual users, the minimum sample size of 384 respondents is considered sufficient to test the hypothesis. In this study, samples were the international tourists who departed from the Suvarnabhumi Airports of Thailand in May 2012 and the total number of respondents was 708. Therefore, the sample size of this study was sufficient to test hypotheses because the number of available questionnaires in the study exceeds 708 cut-off values as suggested by Sekaran and Bougie [2010].

4.2. *Data collection*

Survey respondents participated in this study voluntarily. The convenience sampling technique was employed for data sampling. Since the sample of this study has no sampling frame or the list of all international tourists who visited Thailand, the probability sampling method could not be employed for collecting data. Hair *et al.* [2010] stated that convenience sampling technique could be used to select large number of sample members who could provide the required information and are available to participate in the study. In addition, several previous tourism studies which collected data from individuals also employed the convenience sampling technique in their researches [Simon and Gladys, 2011; Tsang *et al.*, 2011]. Thus, convenience sampling technique was the appropriate technique for this study.

Before filling out the survey, participants were given the demonstration of MTG usage and MTG features, and they had a chance to try out the MTG. A total of 708 responses were collected and included in the final analysis. The male respondents were more than half of respondents (54.5%). The majority was between 11 and 40 years old (57.0%). With regard to marital status, the data shows that 56.2% of respondents were married and in relationship, while 43.8% of the respondents were single, widowed, or divorced. From the data, 47.0% of respondents had an annual income greater than US$40,000, while others had annual income less than US$40,001 (53.0%) (see Table 2). In terms of region of origin, the majority of the respondents were Asian (40.1%), while the other 59.9% of them came from Europe, America, Oceania, Africa, and Middle East.

5. Data Analysis and Results

5.1. *Structural model*

The research model in this study was examined using WarpPLS 3.0 and the results of path analysis as shown in Figure 2. The R^2 value of research model indicates that the model variance

Table 2. The demographic data of respondents.

Profile	Item	Frequency	Percentage
Gender	Male	388	54.5
	Female	320	45.2
Age	Below 11 years	0	0
	11–20 years	36	5.1
	21–30 years	336	47.5
	31–40 years	158	22.3
	41–50 years	74	10.5
	Above 50 years	104	14.7
Marital Status	Single	282	39.8
	Married	314	44.4
	Widowed	2	0.3
	Divorced	26	3.7
	Relationship	84	11.8
Annual Income (US$x1,000)	Below $20,000	176	26.8
	$20,001–$30,000	122	18.6
	$30,001–$40,000	76	11.6
	$40,001–$50,000	74	11.3
	Above $50,000	208	31.7
Continent	Asia	284	40.1
	Europe	216	30.5
	America	60	8.5
	Oceania	108	15.3
	Africa	15	2.3
	Middle East	24	3.4

explained of intention to use MTG was 47.0%. The results of study show that the relative advantage ($\beta = 0.12$, $p < 0.01$), compatibility ($\beta = 0.40$, $p < 0.001$), lack of complexity ($\beta = 0.13$, $p < 0.05$), and trialability ($\beta = 0.14$, $p < 0.001$) positively relate to the international tourists' intention to use MTG. Thus, H1, H2, H3, and H4 are supported. In addition, the role of interaction quality as moderator ($\beta = 0.12$, $p < 0.01$) was found on the relationship between lack of complexity and intention to use MTG.

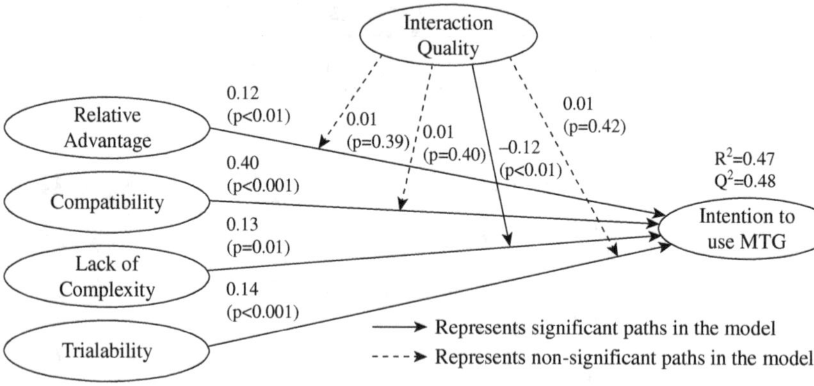

Figure 2. Results of PLS analysis.

It means that the relationship between lack of complexity and intention to use MTG will be stronger when interaction quality is higher as illustrated in Figure 3. Hence, H7 is supported. However, unlike our expectations, the role of interaction quality did not have any impact on three innovation characteristics (relative advantage, compatibility, and trialability). Thus, H5, H6, and H8 are not supported.

According to [Cohen, 1988], the effect size (f^2) values of latent variables could be specified to the small, medium, and large effects by values of 0.02, 0.15 and 0.35, respectively. In this study, the effect sizes of relative advantage ($f^2 = 0.07$), lack of complexity ($f^2 = 0.06$), and trialability ($f^2 = 0.06$) were the small effects in research model. Meanwhile, the effect size of compatibility was medium effect ($f^2 = 0.25$). Furthermore, the interaction quality as moderator also had a small effect ($f^2 = 0.04$).

A value of Q^2 larger than zero indicates that exogenous latent variables have predictive relevance for a particular endogenous latent variable [Chin, 1998]. The Q^2 value for the intention to use MTG in this study was 0.48, which larger than zero. Therefore, our research model was the predictive relevance of the associated PLS path model relationships.

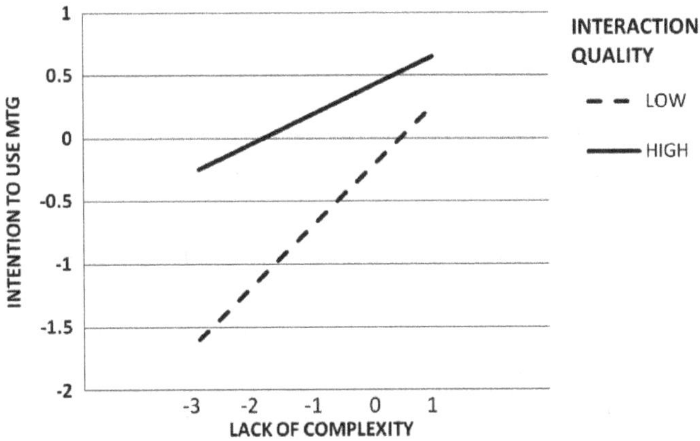

Figure 3. The moderating effect of interaction quality on the relationship between lack of complexity and intention to use MTG.

Overall, our empirical results supported the hypotheses that intention to use MTG was positively associated to innovation characteristics in relative advantage, compatibility, lack of complexity, and trialability. Furthermore, interaction quality could strengthen on the relationship between lack of complexity and intention to use MTG.

6. Discussions and Conclusions

6.1. *Key findings*

This study introduced the moderating variable into IDT to investigate international tourists' intention to use MTG using PLS. Comparing our results with previous IT researches, the study reveals several interesting results. As hypothesized, the study found significant support for the conventional IDT-related constructs such as relative advantage, compatibility, lack of complexity, trialability, and intention. This would imply that the international tourists may have the intention of using MTG if they realize that doing so is useful, compatible with their lifestyle and travel, easy to use and learn. Furthermore, getting MTG on trial version and features demonstration also increase their intention to use MTG.

Relative advantage has a positive effect toward intention in the context of MTG. This finding also supports the concept of TAM, UTAUT, and IDT. Furthermore, this result is consistent with Kim *et al.* [2010] finding in an empirical examination of factors influencing the intention to use mobile payment, and Xue *et al.* [2012] finding in an exploratory study of ageing womens' perception on access to health infomics via a mobile phone-based intervention. Likewise, it confirms the Ayeh *et al.* [2013] finding in predicting the intention to use consumer-generated media for travel planning. This finding implies that higher usefulness will increase the willingness to use MTG. The higher usefulness means that the MTG usage will make getting travel information easier, finding travel guide information quicker, and travel information more effective.

Similarly, the lack of complexity exhibited a positive impact on intention. This finding confirms the result of the other IT contexts [Kim *et al.*, 2010; Oh *et al.*, 2009; Xue *et al.*, 2012]. This result also supports three famous theories and models such as TAM, UTAUT, and IDT. It implies that the high level of ease in terms of usage and learning will increase international tourists' intention to use MTG.

The finding supports our hypothesis that the compatibility has a positive impact on international tourists' intention to use MTG. This implies that higher compatibility will increase international tourist's intention to use MTG. International tourists expect compatibility in terms of fit-into-travel, fit-into-travel style, and suitable for travel guide. The result also is consistent to the other mobile IT contexts [Baird *et al.*, 2011; Lu *et al.*, 2011; Putzer and Park, 2010; Xue *et al.*, 2012]. This finding is very important particularly for international tourists who usually use the MTG in three phases such as pre-trip, during-trip, and post-trip.

In terms of trialability, the result of this study confirms in the direction of relationship between trialability and intention in the context of MTG. This study found the trialability has a positive impact on tourists' intention to use MTG, even though the other contexts in previous studies found a negative impact on users' intention to use IT and mobile IT [Hernandez and Mazzon, 2007; Gerpott, 2011]. It indicates that international tourists expect to receive a demo and experiment before deciding to use the MTG. This result confirms the previous studies that trialability is an important factor of users' intention to use mobile IT [Brown *et al.*, 2003; De Marez *et al.*, 2007] and IT [Askarany *et al.*, 2007; Jung *et al.*, 2012; Ndubisi and Sinti, 2006].

Regarding the moderating role of interaction quality, interaction quality has a role on the relationship between lack of complexity and intention as our expectation. It can imply that the relationship between lack of complexity and intention to use MTG will be stronger when interaction quality is higher. However, interaction quality could not strengthen the relation between intention and the other innovation characteristics (relative advantage, compatibility, and trialability). According to Sharma *et al.* [1981], if interaction of moderating variable (z) and predictor (x) does not have a significance on dependent variable (y) and z relates to x or y, then z cannot be a moderator; it should be antecedent or exogenous. Our result is consistent with the result of study by Tan and Chou [2008], which found the impact of interaction quality on technological compatibility in the context of mobile information and entertainment service. Therefore, the interaction quality may be the antecedent or exogenous for relative advantage, compatibility, and trialability in the context of intention to use MTG. These assumptions could be investigated in future research.

6.2. *Implications of theory and practice*

In terms of theoretical contributions, this study extends the body of knowledge in users' intention to use mobile IT. The study extends and reveals the role of interaction quality on innovation characteristics in IDT. The findings give new insight in a role of interaction quality as a moderating effect on the relationship between lack of complexity and users' intention to use mobile IT, especially the MTG context. Furthermore, the study provides a clear understanding of the innovation characteristics as the determinants of international tourists' intention to use MTG. There are four innovation characteristics (relative advantage, compatibility, lack of complexity, and trialability), which impact on international tourists' intention in the context of MTG. Our study context extends the scope of mobile IT contexts in the users' intention study.

In terms of practical contributions, the study provides important guidance to better understand and help tour organizations in terms of the MTG development and implementation. First, in order to support international tourists' intention, the MTG developers should design and develop based on the four innovation characteristics of MTG such as relative advantage, compatibility, lack of complexity, and trialability. Second, the MTG developers should emphasize the interaction quality when they design and develop the MTG or mobile IT. As a finding of this study, the higher level of interaction quality can increase the lack of complexity influence on users' intention. The high interaction quality means that the menus should be clearly categorized, the title bar should be provided in every screen in order to easily recognize the screen location, and the layout should be consistent to support the users' perception in ease-to-use. Third, the tour operators should provide the trial mobile application versions and demonstrations for international tourists to see the overview of tour contents and features; due to the finding of this study, it evidences that trialability is a key factor of international tourists' intention to use MTG. Fourth, the tour operators should highlight their MTG to target users in terms of the four innovation characteristics (relative advantage, compatibility, lack of complexity, and trialability) via advertisements. This operation will increase the tour operators' success in MTG launching and implementation. Fifth, the MTG should be support tour contents in three travel phases (pre-travel, during travel, and post-travel), in order to support international tourists' intention to use MTG. For instance, supporting trip planning before travel, getting tour information during travel, and sharing trip experiences after travel. Finally, the mobile IT researchers and academicians should include the relative advantage and lack of complexity in their studies, when they desire to study the users' intention in the mobile IT context. Furthermore, they should include and examine the compatibility in their study contexts. For example, fit-to-work style in the mobile workspace, fit-to-travel style in the MTG, and fit-to-lifestyle in the mobile movie ticket. In addition, they should include the trialability when their mobile IT is new for their target users.

6.3. *Limitation and future research*

The limitation of this study is the sample, even though the samples are representative of international tourists in Thailand as the continent, gender, and age of sample proportions in this research are nearly the actual international tourist proportions of Thailand. Therefore, the other samples from different areas or nations should be gathered to confirm and refine. Additionally, the test–retest reliability of instrument also should be evaluated in terms of internal consistency and stability over time in order to assess the intention to use mobile technology in different periods.

Summary, this study developed and tested models that examined the innovation characteristics on MTG. Furthermore, the result demonstrated the interaction quality as new construct in IDT, moderated on relationship between lack of complexity and intention to use. The findings offer new insights on intention to use MTG through assessment of their structure. Thus, our model is significant in predicting intention to use MTG among international tourists.

References

Akter, S., J., Ambra, D. and Ray, P. (2010). Service quality of mHealth platforms: Development and validation of a hierarchical model using PLS. *Electronic Markets*, **20**(3):209–227.

Askarany, D., Smith, M. and Yazdifar, H. (2007). Attributes of innovation and the implementation of managerial tools: an activity-based management technique. *International Journal of Business and Systems Research*, **1**:98–114.

Ayeh, J. K., Au, N. and Law, R. (2013). Predicting the intention to use consumer-generated media for travel planning. *Tourism Management*, **35**:132–143.

Baird, A., North, F. and Raghu, T. (2011). Personal Health Records (PHR) and the future of the physician–patient relationship. *Proceedings of the 2011 Conference*, ACM, Seattle, WA, USA, 8–11 February 2011, pp. 281–288.

Brown, I., Cajee, Z., Davies, D. and Stroebel, S. (2003). Cell phone banking: Predictors of adoption in South Africa-An exploratory study. *International Journal of Information Management*, **23**(5): 381–394.

Carter, L. and Belanger, F. (2004). The influence of perceived characteristics of innovating on e-government adoption. *Electronic Journal of E-government*, **2**(1):11–20.

Chae, M., Kim, J., Kim, H. and Ryu, H. (2002). Information quality for mobile Internet services: A theoretical model with empirical validation. *Electronic Markets,* **12**(1):38–46.

Chin, W. W. (1998). *The PartialLleast Squares Approach for Structural Equation Modeling*. Lawrence Erlbaum Associates Publishers, Mahwah, NJ.

Cohen, J. (1988). *Statistical Power Aanalysis for the Behavioral Sciences*. Lawrence Erlbaum Associates, Hillsdale.

Dai, H. and Palvi, P. C. (2009). Mobile commerce adoption in China and the United States: A cross-cultural study. *ACM SIGMIS Database*, **40**:43–61.

Davis, F. (1989). Perceived usefulness, perceived ease of use, and user acceptance of information technology. *MIS Quarterly*, **13**:319–340.

De Marez, L., Vyncke, P., Berte, K., Schuurman, D. and De Moor, K. (2007). Adopter segments, adoption determinants and mobile marketing. *Journal of Targeting, Measurement and Analysis* for Marketing, **16**(1):78–95.

Flurry Analytic (2012). App session growth by country between Q1 2011 and Q2 2012. Available at http://blog.flurry.com/bid/83261/China-Now-Leads-the-World-in-New-iOS-and-Android-Device-Activations.

Gerpott, T. J. (2011). Attribute perceptions as factors explaining Mobile Internet acceptance of hcellular customers in Germany — An empirical study comparing actual and potential adopters with distinct categories of access appliances. *Expert Systems with Applications*, **38**(3):148–216.

Hair, J. F., Black, W. C., Babin, B. J. and Anderson, R. E. (2010). *Multivariate Data Analysis,* 7th edn. Pearson Prentice Hall, Upper Saddle River, New Jersey.

Hernandez, J. and Mazzon, J. (2007). Adoption of internet banking: Proposition and implementation of an integrated methodology approach. *International Journal of Bank Marketing,* **25**:72–88.

Hsu, C. L., Lu, H. P. and Hsu, H. H. (2007). Adoption of the mobile Internet: An empirical study of multimedia message service (MMS). *Omega*, **35**(6):715–726.

Jung, J., Chang-Olmsted, S., Park, B. and Kim, Y. (2012). Factors affecting e-book reader awareness, interest, and intention to use. *New Media & Society*, **14**:204–224.

Kim, J. S., and Kizildag, M. (2011). M-learning: Next generation hotel training system. *Journal of Hospitality and Tourism Technology*, **2**: 6–33.

Kim, C., Mirusmonov, M. and Lee, I. (2010). An empirical examination of factors influencing the intention to use mobile payment. *Computers in Human Behavior*, **26**:310–322.

Koivumaki, T., Ristola, A. and Kesti, M. (2008). The effects of information quality of mobile information services on user satisfaction and service acceptance — Empirical evidence from Finland. *Behaviour & Information Technology*, **27**(5):375–385.

Kuo, Y. F., Wu, C. M. and Deng, W. J. (2009). The relationships among service quality, perceived value, customer satisfaction, and post-purchase intention in mobile value-added services. *Computers in Human Behavior*, **25**(4):887–896.

Liao, H. L. and Lu, H. P. (2008). The role of experience and innovation characteristics in the adoption and continued use of e-learning websites. *Computers & Education*, **51**(4):1405–1416.

Lin, T. C., Cheng, H. K., Wang. F. S. and Chang, K. J. (2012). A study of online auction sellers' intention to switch platform: The case of Yahoo! Kimo Versus Ruten_eBay. *Decision Sciences*, **43**(2):241–272.

Lu, Y., Yang, S., Chau, P. Y. K. and Cao, Y. (2011). Dynamics between the trust transfer process and intention to use mobile payment services: A cross-environment perspective. *Information and Management*, **48**:393–403.

Mallat, N. (2007). Exploring consumer adoption of mobile payments — A qualitative study. *The Journal of Strategic Information Systems*, **16**(4):413–432.

Moore, G. C. and Benbasat, I. (1991). Development of an instrument to measure the perceptions of adopting an information technology innovation. *Information Systems Research*, **2**(3):192–222.

Ndubisi, N. O. and Sinti, Q. (2006). Consumer attitudes, system's characteristics and internet banking adoption in Malaysia. *Management Research News*, **29**(1):16–27.

Oh, S., Lehto, X. Y. and Park, J. (2009). Travelers' Intent to Use Mobile Technologies as a Function of Effort and Performance Expectancy. *Journal of Hospitality Marketing and Management*, **18**:765–781.

Park, Y. J., Heo, P. S., Rim, M.H. and Park, D. S. (2008). Customer satisfaction index measurement and importance-performance analysis for improvement of the mobile RFID services in Korea. *PICMET '08 - 2008 Portland International Conference on Management of Engineering & Technology*, IEEE, Cape Town, South Africa, 27–31 July 2008, pp. 2657–2665.

Putzer, G. J. and Park, Y. (2010). The effects of innovation factors on smartphone adoption among nurses in community hospitals. *Perspectives in Health Information Management/AHIMA, American Health Information Management Association*, **7**:1–20.

Rogers, E. M. (2003). *Diffusion of Innovations*, 5th edn. The Free Press: NY.

Schierz, P., Schilke, O. and Wirtz, B. (2010). Understanding consumer acceptance of mobile payment services: An empirical analysis. *Electronic Commerce Research and Applications*, **9**:209–216.

Sekaran, U. and Bougie, R. (2010). *Research Methods for Business: A Skill Building Approach*, 5th edn. John Wiley and Sons Ltd, United Kingdom.

Sharma, S., Durand, R. M. and Gur-Arie, O. (1981). Identification and analysis of moderator variables. *Journal of Marketing Research*, **18**:291–300.

Simon, W. C. K. and Gladys, L. F. C. (2011). A study of pre-trip use of travel guidebooks by leisure travelers. *Tourism Management*, **32**: 616–628.

Tan, F. B. and Chou, J. P. C. (2008). The relationship between mobile service quality, perceived technology compatibility, and users' perceived playfulness in the context of mobile information and entertainment services. *International Journal of Human–Computer Interaction*, **24**(7):649–671.

Thong, J. Y. L., Hong, W. and Tam, K. Y. (2002). Understanding user acceptance of digital libraries: what are the roles of interface characteristics, organizational context, and individual differences? *International Journal of Human-Computer Studies*, **57**(3):215–242.

Tsang, N. K. F., Chan, G. K. Y. and Ho, K. K. F. (2011). A holistic approach to understanding the use of travel guidebooks: Pre-, during, and post-trip behavior. *Journal of Travel and Tourism Marketing*, **28**:720–735.

Venkatesh, V., Morris, M. G., Davis, G. B. and Davis, F. D. (2003). User acceptance of information technology: Toward a unified view. *MIS Quarterly*, **27**:425–478.

Wu, I. L., Li, J. Y. and Fu, C. Y. (2011). The adoption of mobile healthcare by hospital's professionals: An integrative perspective. *Decision Support Systems*, **51**:587–596.

Xue, L., Yen, C. C., Chang, L., Chan, H. C., Tai, B. C., Tan, S. B., Duh, H. B. L. and Choolani, M. (2012). An exploratory study of ageing women's perception on access to health informatics via a mobile phone-based intervention. *International Journal of Medical Informatics*, **81**:637–648.

Chapter 10

Emerging Applications: Users with Disabilities

Brenda Mak, Paul Beckman, and Nicole Bohn

1. Introduction

Mobile traffic has been growing rapidly. In 2013, the total mobile subscriptions were 6.7 billion [Ericsson, 2014]. With a world population of 7.15 billion [Census Bureau, 2014], the global mobile penetration was 92%. Mobile technology has allowed people and communities to connect, and has served as a means of social engagement for persons with disabilities.

More than a billion of the world population live with some form of disability. In the United States, about 36 million people, or 12% of the US population, have a disability, including *vision, hearing,* and *mobility* impairments [Disabled World, 2011]. For many persons with disabilities, the mobile phone is an innovation that provides them with many opportunities for development [ITU, 2013]. The use of the mobile phone is instrumental in allowing the independent living of persons with disabilities. Feature phones provide a means of communication for people with disabilities. Mobile devices allow them to engage in social networking. Mobiles devices also provide them with sense of security and assurance of emergency help.

The Convention on the Rights of Persons with Disabilities, which came into force in 2008, [ITU, 2012], represents a commitment of the international community and the United Nations to protect the rights of persons with disabilities in all aspects of development. *The Convention recognizes that disability results from the interaction between persons with impairments and the barriers (both attitudinal and environmental) that hinder their full and effective participation in society on an equal basis with others.*

The UN Convention on the Rights of Persons with Disabilities represents an international milestone acknowledging the shift in attitudes and approaches to persons with disabilities. Of particular importance is the accessibility of information and communications technology (ICT). To contribute to the efforts leading to the formulation of a development agenda, the ICT Opportunity for a Disability-Inclusive Development Framework was produced in 2013 by the United Nations, International Telecommunication Union (ITU), Microsoft and the International Disability Alliance. This framework identifies best practices and challenges to which ICTs and mobile devices can enable and accelerate social and economic inclusion of persons with disabilities. The Framework outlined the importance of research and development efforts to understand the needs and concerns of users with disabilities for incorporating universal design in communication devices. The government was also recommended to play an active role in promoting communication technologies adapted to the needs of persons with disabilities through revision of disability legislations. In light of these regulatory changes, research is required for the industry and practitioners to understand the needs of the disability market segments, so as to find out the elements affecting the usefulness and satisfaction of the mobile phone to users with disabilities.

2. Research Problem

For most persons with disabilities, the mobile phone is an innovation. Studies have been conducted on the adoption of innovations. Rogers [1995] suggested an individual's decision to adopt an innovation is a process that occurs over time. An innovation is adopted more easily if the number of relative advantages is high. Davis *et al.* [1989] proposed the technology acceptance model (TAM), suggesting that a person's attitude toward technology will affect his or her satisfaction and intention to use it, while this attitude toward technology is in turn affected by the perceived usefulness and perceived ease of use of the technology.

The mobile phone serves as the most frequently used media and communication tool for social participation for persons with disabilities [ITU, 2013]. Social function is the *first* important aspect of the perceived usefulness of the mobile phone. For persons with disabilities, improvement in mobile technology has enabled people with disabilities to participate in social activities anywhere, anytime [Burton, 2011]. A second important aspect of perceived usefulness is the emergency help the mobile phone provides. The mobile devices can enable independent living by providing emergency services to the person with disabilities. Family members, agencies, or friends can be contacted with the emergency call placed by a subscriber. A third aspect of perceived usefulness of the mobile phone is the sense of security provided. Researchers have found that mobile phones provide sense of security for weak and frail individuals, allowing them to be reached anytime and anywhere [Kurniawan *et al.*, 2006; Kurniawan, 2008].

The purpose of this research is to study the relationships between the perceived usefulness of mobile phones and user satisfaction for persons with disabilities. Perceived usefulness is assessed as a latent construct measured in terms of perceived social function, emergency help, and sense of security provided by the mobile phone. To investigate these relationships, a survey was conducted among mobile phone users with disabilities in the San Francisco Bay Area.

From an academic point of view, this study enriches the understanding of mobile phone's perceived usefulness and their relationship to user satisfaction, and further the insight on how the TAM can be applied to understand the needs of users with disabilities. This understanding is particularly important as the ICT Opportunity for a Disability-Inclusive Development

Framework by the ITU and the United Nations calls for research to identify best practices and challenges to which mobile devices can enable social and economic inclusion of persons with disabilities. From a practical point of view, the study helps to understand the elements involved in perceived usefulness for people with disabilities. Providers and manufacturers can design features to enhance user satisfaction, supporting the implementation of the ICT Opportunity for a Disability-Inclusive Development Framework to meet the needs of users with disabilities.

3. A Model for Mobile Phone Perceived Usefulness & Satisfaction

3.1. *Perceived usefulness factors: Social function*

Perceived social function is an important aspect of the perceived usefulness of the mobile phone for users with disabilities. ICT enables and accelerates the social economic inclusion of persons with disabilities [ITU, 2013]. Of particular importance is the mobile device which allows instantaneous access to all aspects of society and development. Opportunities for social participation also include social networking, news access, online interest groups, video, audio and text communication, cloud-based sharing, and media interaction. For persons with disabilities, these services and content are made further accessible through both computer-based and web-based accessibility applications such as screen readers, speech recognition, video communication (for sign language communication and video relay interpretation), voice-to-text services (open and closed captioning, both real-time and embedded), and visual assistance.

Past research has demonstrated that social function is an important aspect of the perceived usefulness of the mobile phone. Aoki and Downes [2003] found that even for users who might have adopted cell phones for security- or job-related reasons, the mobile phones will be increasingly used for social interactions. In investigating the psychological influences of mobile phone use, Walsh *et al.* [2007] found that social factors were the strongest reason for mobile phone use. In comparison to the Internet, Humphreys [2008] found that the mobile phone is able to create a more instantaneous public space for social relations and interaction when the mobile social network system Dodgeball was examined. Fortunati *et al.* [2010] found that the mobile phone is the most important instrument for social communication when compared with the Internet and other traditional media.

Research has found that social network externalities affect perceived usefulness and user satisfaction. In a study of 223 users, Zhou and Lu [2011] found that satisfaction and loyalty increase if users perceive most of their friends and classmates are using the mobile instant messaging service. Thus, social function is one important aspect of perceived usefulness of the mobile phone and users adopt mobile instant messaging to interact with peers in their social circle. In a study of 402 Facebook users, Lin and Lu [2011] found that enjoyment, number of peers, and perceived usefulness were the most influential factors affecting the continued use of the social networking site. When the user believes it upgrades the user's connecting with others, the user's continued intention to use the site increases.

In analyzing the sociological impact of the mobile phone, Geser [2004] suggests that the mobile phone is instrumental in satisfying the communication needs of individuals. The mobile phone allows individuals to engage in communication free from constraints of physical proximity and spatial immobility. Geser also suggests that the mobile phone bridges the gaps of communication between different social classes, freeing individuals from their immediate social surroundings by empowering them to fulfill many material and psychological needs without relating to any others in their vicinity. The mobile phone allows individuals to defend a minimal private social space and keep them connected to their close kin or friends regardless of their locations.

Thus, perceived social function is a key aspect of the perceived usefulness of the mobile phone. We posit that the latent perceived usefulness of the mobile phone construct can be manifested in terms of the social function the mobile phone provides to the user.

3.2. *Perceived usefulness factors: Emergency help*

The provision of emergency help is an important aspect of perceived usefulness of the mobile phone. For persons with disabilities, having a mobile device increases independent living because of the wide range of emergency services that can be accessed. In addition, the fact that the mobile phone is portable and easily carried by a user allows easy access to emergency services immediately at the time of need and from anywhere in the network [ITU, 2013].

Various researchers have studied the emergency use of mobile phones. Tennakoon and Taras [2012] conducted interviews to study how mobile phones are used during emergencies. They found out that the mobile phone provides a mode of communication to reach help anywhere, anytime and acts as an emergency contact medium. They also found that the mobile phone serves as a companion when no one else was there, and reduces the threat of an attack. When there are imminent dangers in the vicinity, for example, in the case of a bomb blast emergency, the mobile phone allows communication of reassuring words to friends and families, and provides an informal network of information dissemination and news alert.

Holzinger *et al.* [2007] conducted experiments to improve the interfaces to minimize the amount which needs to be memorized, so that the connection to emergency services is simplified in case of emergency. Ketabdar and Polzehl [2009] studied how mobile phones can be designed to monitor the physical activities of users and detect unexpected emergency situations such as a sudden fall or accident. Upon detection of such an event, the mobile phone can inform a designated center by automatically calling or sending messages about the incident and its location. Such an application can be particularly useful for elderly people or people with physical and movement disabilities.

In a study of the use of mobile phone by people aged 60 years and over, Kurniawan [2008] found that the most common response for using mobile phones was for emergency. In addition, respondents also indicated after acquiring mobile phones, they feel safer, more confident to go out by themselves, not afraid of getting lost, and know they can always call somebody for help if they are in trouble. As suggested by the ITU [2013], mobile phones enable independent living by ensuring emergency services are readily available for the elderly and persons with disabilities. Thus, emergency help is a key aspect of the perceived usefulness of the mobile phone. We posit that the latent perceived usefulness of the mobile phone construct can be manifested in terms of the emergency help the mobile phone provides to the user.

3.3. *Perceived usefulness factors: Sense of security*

A third aspect of its perceived usefulness is the sense of security provided by the mobile phone. ITU [2013], together with UNESCO, Microsoft, G3ict, International Disability Alliance, and Telecentre.org Foundation, conducted a consultation with experts including government, academic institutions, organizations of persons with disabilities, private sectors, and international organizations. They found that the mobile phone is the most valued communication technologies in bringing about independent living for persons with disabilities, improving their access to various activities with peace of mind, and providing them services at the time of need.

In a focus group study with older women, Kurniawan *et al.* [2006] found that the mobile phone provides sense of security for them, as individuals can be reached anytime, anywhere. In a follow-up study with panel experts and older men,

Kurniawan [2008] found that the mobile phone enables frail older persons to engage in active social life, and provides them with a sense of safety and security. The sense of security is one of the primary reasons for their use of their mobile phones. He found that respondents had favorable responses toward the perceived sense of safety associated with mobile phones, and respondents thought the mobile phone had a positive role in facilitating their feelings of security.

Yuan [2007] found that the mobile phone brings peace of mind to the user with crime alerts. Aoki and Downes [2003] studied motivations for acquiring mobile phones using focus groups and a survey, and found that sense of security was an initial motivator, and respondents felt less secure when deprived of their phones.

Tennakoon and Taras [2012] studied usage patterns and user sentiments about mobile phone technologies in Canada and Sri Lanka. They found that respondents in both countries demonstrated a heightened sense of security with mobile phone use, with females showing significantly higher perception of security than males. When individuals had higher perceived usefulness of the mobile devices, they also had a higher sense of security associated with mobile phone use. Certain emotions and cognitive mechanisms were at work to bring about this sense of security, as respondents indicated that the mere presence of the mobile device seemed to bring comfort and peace of mind.

Thus, the sense of security provided by the mobile phone is one primary factor of its perceived usefulness. We posit that the latent perceived usefulness of the mobile phone construct can be manifested in terms of the sense of security the mobile phone provides to the user.

3.4. *Relationship between perceived usefulness and landline possession*

As suggested by the TAM, Davis *et al.* [1989] perceived that usefulness of a new technology is affected by external variables.

Some of the external variables included demographic factors such as age, education, income, race, gender, and marital status [Porter and Donthu, 2006; Rice and Katz, 2003]. One of the most important external variables is possession of landline.

In developing a sociological theory of the mobile phone, Geser [2004] noted that the need for communication has led to the invention of the phone. Geser suggested that while the landline phone required individuals to stay at specific places, the mobile phone was an innovation that makes communication compatible with spatial mobility. The landline phone, in history, had been adopted for functional purposes and later used for social purposes [Palen *et al.*, 2000]. Palen *et al.* [2000] interviewed new mobile phone users and found that similar to the landline phone, the early adoption of the mobile phones was for functional purposes, but eventually the mobile phones were used for social reasons. Mobile phones are part of the social world, and can be used at times when and at places where landlines were not normally used in the past.

As suggested by previous researchers [Palen *et al.*, 2000; Wei, 2006], individuals adopting the mobile phones would perceive higher values of the mobile phone. Thus, landline possession is a significant factor affecting the latent variable of perceived usefulness and its relationship between perceived usefulness and satisfaction. In this research, we study how landline possession will affect perceived usefulness of mobile phone users. We posit the following hypothesis:

H1: Individuals with landline possession would have higher perceived usefulness of the mobile phone.

3.5. *Relationship between perceived usefulness and satisfaction*

Various researchers have studied the effect of perceived usefulness on user satisfaction of mobile phones. Lai *et al.* [2009] conducted a survey of customers of a mobile communications company and found that the perceived value of mobile phones has a positive impact on user satisfaction. Chiu *et al.* [2005] found that the perceived usefulness of e-learning service programs is positively related to user satisfaction. McDougall and Levesque [2000] found that core service quality and perceived values are the important drivers of customer satisfaction, lowering intentions to switch to competitors. Research has also found that the feelings of being accepted by others and making a good impression on other people are determinants of customers' satisfaction with the mobile messaging service [Deng *et al.*, 2010]. Therefore, we posit that the latent construct of mobile phone perceived usefulness have important satisfaction impacts.

H2: The mobile phone perceived usefulness is *positively* related to user satisfaction with mobile phones. Higher perceived usefulness leads to higher user satisfaction.

The focus of this research is to explore the factors of perceived usefulness of the mobile phone for persons with disabilities, and how these factors affect user satisfaction. Specifically, we posit that social function, emergency help, and sense of security are three aspects of mobile phone's perceived usefulness to persons with disabilities, and *perceived usefulness affects perceived satisfaction positively.*

4. Methodology

4.1. *Research instrument*

The initial survey instrument was constructed based on literature review and consultation with persons with disabilities. The scales were patterned after existing instruments, specifically, the studies of Turner *et al.* [2008], Aoki and Downes [2003], and Wei [2006] on social function and sense of security of the mobile phone. The instrument was refined by holding interviews with mobile phone users with disabilities. The questionnaire was

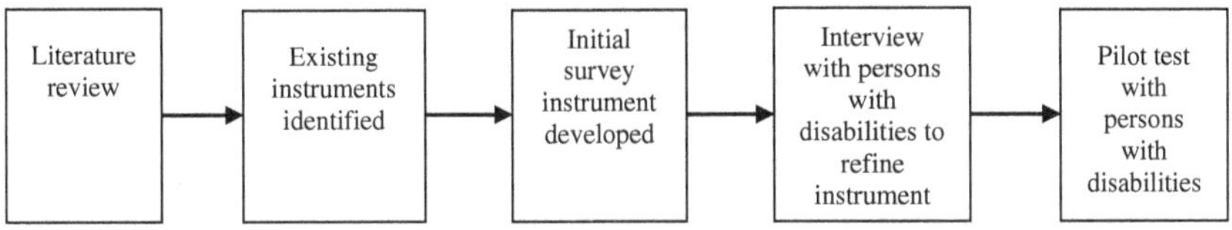

Figure 1. Steps of developing the survey instrument.

subsequently refined based on the feedback provided. A pilot study was conducted with a sample of two students with disabilities. The study was conducted in labs with assistive technologies so that the students could complete the survey on workstations with adaptive technologies. They were asked to fill out the questionnaire online, and to evaluate on the accessibility and usability of the survey. The pilot study helped to test and validate the scales of the instrument adopted for the final survey. Responses to the pilot test provided feedback on item wording and format, and helped to further refine the survey questionnaire. Figure 1 shows the steps of developing the survey instrument.

Respondent perceptions were measured using a Likert scale of 6 points, with 1 for strongly agree, 2 for moderately agree, 3 for slightly agree, 4 slightly disagree, 5 for moderately disagree, and 6 for strongly disagree. The even number of response categories was adopted to force respondents to be more discriminating and thoughtful in selecting their responses, and to ensure that accurate responses were collected [Losby and Wetmore, 2012]. Table 1 displays the items used in the questionnaire.

4.2. *The sample*

The sample was obtained from a diverse population of students with disabilities served by the Disability Programs and Resource Center (DPRC) at a university in the San Francisco Bay Area. One of the top 15 universities for campus diversity according to U.S. News and World Report 2009 rankings, the university's students

Table 1. Items within the questionnaire for mobile phone's perceived usefulness and satisfaction.

Variable	Item	Question	Reliability: Corrected item total correlation	Factor loading
Social function (Cronbach's alpha = 0.91; one factor 79.3% of variance)	SocFun1	My mobile phone keeps me involved with social activities	0.73	0.84
	SocFun2	My mobile phone helps me build my social network	0.82	0.90
	SocFun3	My mobile phone serves many important social functions for me	0.85	0.92
	SocFun4	My mobile phone saves me time to build social relationships	0.81	0.90
Emergency help (Cronbach's alpha = 0.87; one factor 79.9% of variance)	Emergen1	My mobile phone helps me inform classmates and instructors of my emergency situations	0.76	0.89
	Emergen2	My mobile phone is a big help to my friends and family in medical and other emergency needs	0.81	0.92
	Emergen3	My mobile phone reassures me of help in case of emergencies	0.71	0.87

Sense of security (Cronbach's alpha = 0.81; one factor 73.7% of variance)	Securty1	My mobile phone gives me a sense of security and makes me feel safer	0.70	0.87
	Securty2	My mobile phone provides me with psychological comfort	0.67	0.86
	Securty3	My mobile provides me with peace of mind while traveling	0.65	0.84
Perceived satisfaction (Cronbach's alpha = 0.90; one factor 77.3% of variance)	Sat1	I believe that I am satisfied with the service support I get for my mobile phone	0.79	0.88
	Sat2	I believe that I am satisfied with the functions and features of my mobile phone	0.74	0.86
	Sat3	I believe that I am satisfied with the voice clarity I get on my mobile phone	0.74	0.86
	Sat4	I believe that I am satisfied with the reliability of my mobile phone	0.85	0.92

represent a range of backgrounds, interests, and points of view. The questionnaires were distributed to 850 students registered in the DPRC through email. To increase response rate, survey participants had the opportunities to enter into a raffle. The grand prize was an Apple iPad wifi and the second prize was an Apple iPod Touch.

A total of 142 usable questionnaires were collected. The response rate was 16.7%. Hair *et al.* [2010] indicates that for a model with five or less latent constructs, a sample size of 100 is sufficient to produce a nearly normal sampling distribution and to provide adequate power for the estimation of the structural equation model. This sample size is sufficient for estimating the structural equation model in Figure 2.

All of the respondents have had a mobile phone for at least one year, and 68% of the respondents had mobile web access. 55% of them did not have a landline phone. Out of the 142 respondents, 66% were female, 33% were male, and the rest

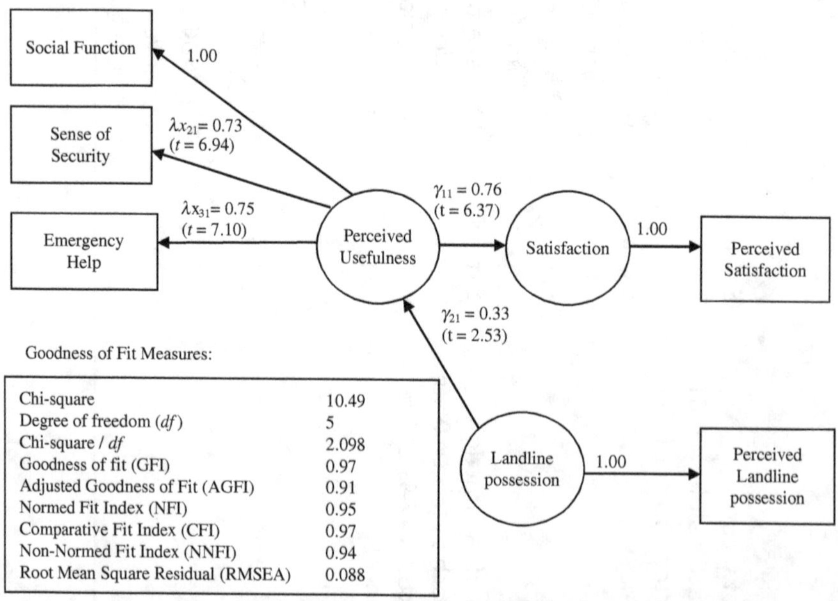

Figure 2. Relationships among perceived usefulness, satisfaction, and landline possession.

preferred not to state their gender. 45% were between the ages 18–25, 19% were between the ages of 26–30, and 34% were above 30, and 2% did not state their age. 86% of the respondents were full-time students while 15% were part-time students. 8% were working full time, 33% were working part time, and 59% were not working. 29% had completed bachelor degrees or above.

The respondents stated their types of disabilities and many of them indicated multiple disabilities. 52% of respondents had learning disabilities, 32% had psychological disabilities, 8% had blindness/visual disabilities, 8% had hard of hearing disabilities, 2% had speech/language disabilities, 20% had mobility/orthopedic disabilities, 23% had health disabilities. 6% had other types of disabilities (epilepsy, traumatic brain injury, hand injuries, lupus, social behavior disabilities, functional disability, autism spectrum disorder, attention deficit hyperactivity disorder). The study was conducted in labs with assistive technologies so that the respondents with disabilities could complete the survey on workstations with adaptive technologies. People who had visual disabilities were able to use screen readers and help at DPRC to complete their survey online.

5. Results

The data were analyzed using LISREL [Jöreskog and Sörbom, 1989], a useful tool for understanding the underlying dimension of a latent variable, such as Perceived Usefulness, and this gives an accurate picture of the simultaneous effect of the factors involved. The effects of demographic variables were examined to assess their relationship with perceived usefulness and satisfaction. These variables include age, gender, number of disabilities, and the length of time the respondent has had the mobile phone, and whether the respondent has landline phone or not. These demographic variables except landline possession had no significant impact on perceived usefulness and satisfaction.

Landline possession has a positive significant impact on perceived usefulness. The final model is given in Figure 2. In the final model, "Perceived Usefulness" is made up of three significant indicators, "Social Function", "Sense of Security", and "Emergency Help". Figure 2 summarizes the parameter estimates and *t*-values for the model constructs. The model parameter λx_{11} is set to unity to define the unit of measurement for the latent variable "Perceived Usefulness". As indicated in Figure 2, λx_{21} ($t = 6.94$) is 0.73 and λx_{31} ($t = 7.10$) is 0.75; both are significant at the 0.05 level. This suggests that "Social Function", together with "Sense of Security" and "Emergency Help", are observable measures of the latent variable "Perceived Usefulness".

Since λx_{21} is positive, this shows that perceived usefulness of the mobile phone increases with the sense of security it offers to persons with disabilities. In addition, λx_{31} is positive. This shows the perceived usefulness of the mobile phone increases with the perceived emergency help it offers to persons with disabilities. Since both λx_{21} and λx_{31} are less than unity, this suggests that both "Sense of Security" and "Emergency Help" are less important indicators than "Social Function" in measuring "Perceived Usefulness". In addition, the magnitude of λx_{21} is about the same as λx_{31}, suggesting that "Sense of Security" and "Emergency Help" have about the same importance in measuring "Perceived Usefulness".

The effect of "Perceived Usefulness" on "Satisfaction" is positive significant, γ_{11} is 0.76 ($t = 6.37$), indicating that the higher the perceived usefulness, the more positive is the user satisfaction. *H2 hypothesizes that the mobile phone's perceived usefulness is positively related to user satisfaction with mobile phones. Thus, H2 is accepted.* On the other hand, γ_{21} is 0.33 ($t = 2.53$) and is significant at the 0.05 level. This shows that landline possession has a positive effect on "Perceived Usefulness", and respondents who possess landlines tend to have higher perceived usefulness of mobile phones than those who do not possess landlines.

H1 hypothesizes that individuals with landline possession had higher perceived usefulness of the mobile phone. Thus, H1 is supported. In addition, landline possession has no direct effect on satisfaction (alternative model tested found no significant relationship between landline possession and satisfaction), but landline possession has an indirect positive effect on user satisfaction through perceived usefulness.

Table 2 compares the average values and standard deviations of the indicator variables "Social Function", "Sense of Security", and "Emergency Help" for respondents with and without landline possession. As shown in Table 2, those with landline possession have significantly higher averages for social function, perceived sense of security, and user satisfaction.

The adequacy of the model in Figure 2 is assessed using various measures. In testing structural equation models, the null hypothesis is set up as *a priori* not to be rejected, while the chi-square statistic tests whether the observed data fit the hypothesis of the proposed model, and a smaller chi-square value indicates a better fit. For small sample sizes that might have slightly departed from normality, instead of using the chi-squares, the chi-square per degree of freedom should be used. A ratio of approximately 5 shows a reasonable fit while a ratio between 1 and 2 is an excellent fit [Hu and Bentler, 1999; Hoe, 2008]. The ratio of the model in Figure 2 is 2.098 (chi-square = 10.49 with 5 degrees of freedom), indicating a good fit.

Other measures of fit include the goodness of fit index (GFI) and normed fit index (NFI). Both the GFI and NFI are always between 0 and 1, with 1 indicating a perfect fit while any value above 0.9 suggests a good fit. The model has a GFI of 0.97 and a NFI of 0.95. This shows a good fit. The adjusted goodness of fit (AGFI) is 0.91. This again shows a good fit. Similarly, the non-normed fit index (NNFI) and the comparative fit index (CFI) are two additional measures ranging from 0 to 1, where values greater than 0.9 represent a good model fit. The NNFI and CFI for

Table 2. Averages and standard deviations of measures for groups with and without landline possession.

Variables	Averages (standard deviations) for respondents without landline possession	Averages (standard deviations) for respondents with landline possession	*t*-statistics	Are the averages significant different at 0.05 significance level?
Social function	2.093 (1.162)	2.712 (1.307)	3.1465	Significant ($p = 0.002$)
Emergency help	1.944 (1.055)	2.022 (1.287)	0.9419	Not significant
Sense of security	2.124 (1.216)	2.452 (1.039)	1.983	Significant ($p = 0.049$)
Perceived satisfaction	2.436 (1.330)	2.847 (1.316)	2.107	Significant ($p = 0.037$)

the model are 0.94 and 0.97, respectively. Finally, the structural equation model was assessed using the root mean square error of approximation (RMSEA) and the standardized root mean square residual (SRMR). The RMSEA is a measure of the discrepancy per degree of freedom for the model. A RMSEA value of 0.05 or less indicates a good model fit while a value of 0.08 or below indicates a reasonable fit. The SRMR is an absolute fix index sensitive to misspecification, and a value of 0.05 or below indicates a good fit. The model has a RMSEA of 0.088 and a SRMR of 0.045, indicating a good model fit. Thus, GFI, AGFI, NFI, NNFI, CFI, SRMR, and RMSEA all indicate that the model has a good fit.

6. Discussion and Future Research

This study examines the factors of "Perceived Usefulness" of mobile phones for persons with disabilities. The results indicate that "Social Function", "Emergency Help", and "Sense of Security" are observable aspects of "Perceived Usefulness". To improve perceived usefulness, providers need to develop better features and functionalities to enhance the role of the mobile phone in providing social function, emergency help, and sense of security to persons with disabilities.

In this study, LISREL was applied to analyze the underlying relationships between perceived usefulness and satisfaction. The result of this study is consistent with results from past research efforts on the TAM and has demonstrated a positive relationship between mobile phone's perceived usefulness and user satisfaction for persons with disabilities. An interesting observation is that the possession of landlines has a positive impact on perceived usefulness. It may be because respondents with landlines feel that the mobile phone, when compared to the landline phone, is more instrumental in serving social functions and in providing sense of security. Our result is consistent with research results which showed that social network has a strong positive effect on mobile user behavior [Ghose and

Han, 2011]. Our result also corroborates with the ones obtained by ITU [ITU, 2013], showing that the mobile phone is instrumental in allowing independent living for persons with disabilities, providing them with sense of security and emergency help, and serving as avenues for social interaction.

Limitations of the respondents in this study are mostly learning disabilities and/or psychological disabilities. Another limitation of this research is that this study focuses on user satisfaction as the effect variable. Future research may include conducting surveys on mobile phone usage with persons with other types of disabilities. Future research may also assess the effect of mobile phone's perceived usefulness on the health and well-being for persons of disabilities.

References

Aoki, K. and Downes, E. (2003). An analysis of young people's use of attitudes toward cell phones. *Telematics and Informatics*, **20**(4): 349–364.

Burton, D. (2011). Cell phone access: The current state of cell phone accessibility. *Access World*, **12**(6). Available at http://www.afb.org/afbpress/pub.asp?DocID=aw120602.

Census Bureau (2014). U.S. Census Bureau — World POPClock Projection. Available at http://www.census.gov/popclock/.

Chiu, C., Hsu, M., Sun, S., Lin, T. and Sun, P. (2005). Usability, quality, value and e-learning continuance decisions. *Computers & Education*, **45**(1):399–416.

Davis, F. D., Bagozzi, R. P. and Warshaw, P. R. (1989). User acceptance of computer technology: A comparison of two theoretical models. *Management Science*, **35**(8):982–1003.

Deng, Z., Lu, Y., Wei, K. K. and Zhang, J. (2010). Understanding customer satisfaction and loyalty: An empirical study of mobile instant messages in China. *International Journal of Information Management*, **30**(4):289–300.

Disabled World (2011). *Latest US Disability Statistics and Facts, Information*. Available at http://www.disabled-world.com/disability/statistics/census-figures.php, Published 26 July 2011.

Ericsson (2014). Ericsson mobility report. Available at http://www.ericsson.com/ericsson-mobility-report.

Fortunati, L., Manganelli, A., Law, P. and Yang, S. (2010). The mobile phone use in Mainland China: Some insights from an exploratory study in Beijing. *Telematics and Informatics*, **27**(4):404–417.

Geser, H. (2004). Towards a sociological theory of the mobile phone, University of Zurich, Working Paper. Available at http://socio.ch/mobile/t_geser1.pdf.

Ghose, A. and Han, S. P. (2011). An empirical analysis of user content generation and usage behavior on the mobile Internet. *Management Science*, **57**(9):1671–1691.

Hair, J. F., Black, W. C., Babin, B. J. and Anderson, R. E. (2010). *Multivariate Data Analysis*. Prentice Hall, Englewood Cliffs, NJ.

Hoe, S. (2008). Issues and procedures in adopting structural equation modeling technique. *Journal of Applied Quantitative Methods*, **3**(1):76–83.

Holzinger, A. Searle, G. and Nischelwitzer, A. (2007). On some aspects of improving mobile applications for the elderly. *Universal Access in HCI, Part I, HCII 2007*, Ed. C. Stephanidis. LNCS 4554, pp. 923–932, Springer-Verlag, Berlin, Heidelberg.

Hu, L. and Bentler, P. M. (1999). Cutoff criteria for fit indices in covariance structure analysis: Conventional criteria versus new alternatives. *Structural Equation Modeling*, **6**(1):1–55.

Humphreys, L. (2008). Mobile social networks and social practice: A case study of Dodgeball. *Journal of Computer-Mediated Communication*, **13**(1):341–360.

ITU (2012). Making mobile phones and services accessible for persons with disabilities, *A Joint Report of ITU — The International Telecommunication Union and G3ict — The Global Initiative for Inclusive ICTs*. Available at http://www.itu.int/ITU-D/sis/PwDs/Documents/Mobile_Repost.pdf.

ITU (2013). *The ICT* opportunity for a disability-inclusive development framework, *Synthesis Report of the ICT Consultation in Support of the High Level Meeting on Disability and Development of the Sixty-Eighth Session of the United Nations General Assembly*, Broadband Commission for Digital Development, G3ict, IDA, ITU, Microsoft, Telecentre.org Foundation & UNESCO, International Telecommunications Union, Geneva, Switzerland.

Jöreskog, K. G. and Söbom, D. (1989). *LISREL 7: A Guide to the Program and Applications*, 2nd edn. SPSS Inc., Chicago.

Ketabdar, H. and Polzehl, T. (2009). Fall and emergency detection with mobile phones. In *Proceedings of the 11th International ACM SIGACCESS Conference on Computers and Accessibility, ASSETS 2009*, 25–28 October 2009, Pittsburgh, Pennsylvania, USA.

Kurniawan, S. (2008). Older people and mobile phones: A multi-method investigation. *International Journal of Human-Computer Studies*, **66**(12):889–901.

Kurniawan, S. H., Mahmud, M. and Nugroho, Y. (2006). A study of the use of mobile phones by older persons. In *Extended Abstracts of CHI'06 on Human Factors in Computing Systems (CHI EA '06)*, 22–27 April 2006, Montreal, QC, Canada, pp. 989–994.

Lai, F., Griffin, M. and Babin, B. (2009). How quality, value, image, and satisfaction create loyalty at a Chinese telecom. *Journal of Business Research*, **62**(10):980–986.

Lin, K. Y. and Lu, H. P. (2011). Why people use social networking sites: An empirical study integrating network externalities and motivation theory. *Computers in Human Behavior*, **27**(3):1152–1161.

Losby, J. and Wetmore, A. (2012). Using Likert scales in evaluation survey work — Centers for disease control. Available at http://www.cdc.gov/dhdsp/pubs/docs/CB_February_14_2012.pdf.

McDougall, G. and Levesque, T. (2000). Customer satisfaction with services: Putting perceived value into the equation. *Journal of Services Marketing*, **14**(5):392–410.

Palen, L., Salzman, M. and Youngs, E. (2000). Going wireless: Behavior & practice of new mobile phone users, In *Proceedings of the 2000 ACM Conference on Computer Supported Cooperative Work, CSCW 2000*, 2–6 December 2010, Philadelphia, Pennysylvania, pp. 201–210.

Porter, C. E. and Donthu, N. (2006). Using the technology acceptance model to explain how attitudes determine Internet usage: The role of perceived access barriers and demographics. *Journal of Business Research*, **59**(9):999–1007.

Rice, R. E. and Katz, J. E. (2003). Comparing internet and mobile phone usage: Digital divides of usage, adoption, and dropouts. *Telecommunications Policy*, **27**(8/9):597–623.

Rogers, E. M. (1995). *Diffusion of Innovations*, 4th edn. Free Press, NY.

Tennakoon, K. L. and Taras, D. G. (2012). The relationship between cell phone use and sense of security: A two-nation study. *Security Journal*, **25**(4):291–308.

Turner, M., Love, S. and Howell, M. (2008). Understanding emotions experienced when using a mobile phone in public: The social usability of mobile (cellular) telephones. *Telematics and Informatics,* **25**(3):201–215.

Walsh, S. P. White, K. M. and Young, R. M. (2007). Young and connected: Psychological influences of mobile phone use amongst Australian youth. In *Proceedings Mobile Media 2007*, Eds. G. Goggin and L. Hjorth. University of Sydney, pp. 125–134, Available at http: //eprints.qut.edu.au/9753/.

Wei, R. (2006). Motivations for using the mobile phone for mass communications and entertainment. *Telematics and Informatics*, **25**(1): 36–46.

Yuan, L. (2007) Murder, she texted: Wireless messaging used to fight crime. *Wall Street Journal*, p. B.1.

Zhou, T. and Lu, Y. B. (2011). Examining mobile instant messaging user loyalty from the perspectives of network externalities and flow experience. *Computers in Human Behavior*, **27**(2):883–889.

Chapter 11

Emerging Applications: Mobile Phones in Public Places

Brenda Mak, Robert C. Nickerson, and Henri Isaac

1. Introduction

1.1. *The use of mobile phones across countries*

The mobile phone is the most widely spread technology with a fast diffusion rate [Kim, 2005]. Mobile phones have brought about improvements in the quality of life [Choi *et al.*, 2007] and new commercial relationships such as mobile banking [Ratten and Ratten, 2007] and mobile customer relationship management [Moedritscher and Mussnig, 2005], offering customers mobile messages with incentives for business and providing customers with instant access to service information. The mobile phone has been in continual incremental innovation [Hacklin *et al.*, 2005] to meet changing market and customer needs.

In 1978, NTT Docomo launched the first commercial mobile phone service in Japan. Since then, the total number of mobile phone subscriptions in the world had reached 3 billion by July 2007 [Green, 2007]. In the United States, the number of mobile phones reached 233 million in 2006 [Central Intelligence Agency, 2007]. In the same year, 466 million mobile phones were in use in the 27-country European Union, and 461 million were used in China [Central Intelligence Agency, 2007]. Mobile phone penetration varies by country, with some countries having no mobile phones and over 30 countries having a penetration rate of more than 100% [Wallace, 2006].

Mobile service adoption varies in different countries. Asia, and especially Japan, leads with NTT DoCoMo's i-Mode. This service has garnered 45 million subscribers since its first introduction in 1999, although only about 7 million of these are outside of Japan and none are in the US [NTT DoCoMo, 2008]. South Korea is another Asian country with significant mobile services [Shim, 2005]. Europe follows Asia in mobile services, although it leads the US.

1.2. *The case of NTT docomo*

In 1996, Koji Oboshi, president of NTT DoCoMo, foresaw that the demand for new mobile phones would soon peak. Mobile phones with new capabilities and services would need to be developed so consumers would trade in their old mobile for a new, improved one [Bradley and Sandoval, 2002]. Koji saw the future of wireless lay in non-voice, data, and communications. In 1997, NTT Docomo worked on a new service called i-mode (information mode, a 2G/3G mobile internet platform) to offer mobile internet service to customers over their mobile phones [Kodama, 2002]. NTT Docomo started the i-mode service in the beginning of 1999. By collaborating with internet service providers [Jonason *et al.*, 2001] and companies such as Sony Computer Entertainment and Sun Microsystems, improvements and new features were made to the phones [Kodama, 2002]. Mobile phones with i-mode had attracted 40 million users in Japan by the end of January 2004. However, the diffusion of i-mode overseas has been slow. Hung and Yeh [2007] suggested that different competitive climates overseas was part of the reason.

1.3. *Research problem*

Another reason for the different diffusion rates in different countries may be the difference in the regulatory and cultural environments. According to Rogers [1995], an individual's decision to adopt an innovation is not instantaneous but a process that occurs over time. An innovation is more readily accepted if its perceived adoption risk is low, and these perceived risks are related to its relative advantage, compatibility, complexity, trialability, and observability. The technology acceptance model (TAM) was proposed by Davis *et al.* [1989] to model how users come to accept and use a technology. An individual's actual use of a new technology is affected by his or her behavioral intention to use, which is in turn affected by the individual's attitude towards the new technology. This attitude to new technology is affected by two key variables: perceived usefulness and perceived ease of use of the technology, both of which can be affected by external variables. Some of these external variables include psychographic factors such as user involvement [Jackson *et al.*, 1997], self-efficacy [Davis *et al.*, 1989], or demographic factors such as education [Agarwal and Prasad, 1999], age [Yang, 2005], gender [Venkatesh and Morris, 2000], and experience [Kim, 2008]. The model was later extended to include the effect of social influence [Venkatesh and Davis, 2000; Venkatesh *et al.*, 2003], and social norm is an important element in affecting user acceptance.

In this research, we will analyze the impact of demographic factors on attitude to accept mobile phones socially, and how this attitude in turn affects the usage of mobile phones. The attitude of users toward the social acceptance of mobile phones is an important compatibility factor [Rogers, 1995]. It is also an indication of the influence of social norm on the usage of mobile phone. The social acceptance of mobile phones can be culturally specific and differ across countries. In addition to age, gender, education, experience, we also analyze the effect of country on attitude and usage.

The purpose of this study is to explore the effect of these demographic factors on the attitudes toward the social acceptance of mobile phones, and how the attitudes in turn affect actual usage of mobile phones. The attitudes toward social acceptance is examined in terms of the acceptance of mobile phone use in public places and social situations, where the prohibition of use could be the result of government regulations, a specific business's rules, or local customs. As mobile phones may be used for voice and text messaging in public, we investigate how the attitudes toward the acceptance of mobile phone use in public places, as a latent construct, might be observed in terms of individual attitudes toward the acceptance of voice and text mobile communication in public places. The perceived user attitudes toward acceptance are measured in two aspects: (i) attitudes toward the acceptance of mobile phone voice messaging in public places and (ii) attitudes toward the acceptance of mobile phone text messaging in public places.

To assess these relationships, a survey was conducted among university students in the United States, France, Turkey, Italy, and Finland. From an academic point of view, this study enriches our understanding of mobile phone attitudes and usage behavior, and their relationship to demographic factors. From a practical point of view, the study helps us to understand the needs and perception of each group of users. Providers might design mobile phone plans, promotion, and marketing suitable to the needs of different user groups. In the following sections, we review the literature on mobile phone use and develop our model, describe the methodology of the study, present the results and limitations of our study, discuss the implications of the findings, and suggest possible directions for future research.

2. Theories on Acceptance of New Technology

In 1962, Rogers introduced the theory on the diffusion of innovation [Rogers, 1962]. *Diffusion of Innovations*, Glencoe: Free Press). Here, diffusion is described as "a process by which an innovation is communicated through certain channels over time among the members of a social system." In this process, the individual "passes from gaining initial knowledge of the innovation, to forming an attitude toward the innovation, to making a decision to adopt or reject, to implementation of the new idea, and to the confirmation of this decision." During this process, the new technology is evaluated against various criteria or attributes, in order to reduce uncertainties involved in replacing the existing one with a new one. Some of these uncertainties may arise related to the perceptions of the innovation's relative advantage, compatibility, complexity, trialability, and observability. According to Rogers, peoples' attitude toward a new technology is a key element in its diffusion.

The TAM was proposed by Davis *et al.* [1989] to model the factors affecting a user's acceptance of a new technology. TAM is an extension of Ajzen and Fishbein's theory of reasoned action (TRA) that is used to explain and predict the behavioral intention and actual behavior of people. In TRA, an individual's actual behavior is affected by his or her behavioral intention, which is in turn affected by the individual's attitude toward the behavior, as well as the individual's subjective norm. The individual's subjective norm is affected by his or her normative beliefs and motivation to comply, while the individual's attitude toward behavior is affected by his or her beliefs and evaluations.

Similar to TRA, in the TAM, an individual's actual use of a new technology is also affected by his or her behavioral intention to use, which is in turn affected by the individual's attitude towards the new technology. However, this attitude to new technology is affected by two key variables: perceived usefulness and perceived ease of use of the technology, both of which can be affected by external variables.

Some of the external variables included demographic factors. For example, Agarwal and Prasad [1999] found that age, gender, education, experience directly or indirectly affect attitudes, usage, and satisfaction of information systems. Porter and Donthu [2006] developed and tested an extended version of the TAM to consider demographic factors. They found that age, education, income and race are associated differentially with beliefs about the Internet, and that these beliefs influence a consumer's attitude toward use of the Internet. Yang [2005] explored how Singaporeans are influenced to adopt the M-commerce. TAM was applied to examine factors affecting Singaporeans' attitudes toward this emerging mobile technology and applications. It was also found consumer innovativeness, past adoption behavior, technology cluster adoption, age, and gender affect their adoption behavior. Results from multiple regression analyses further reveal that male respondents tend to perceived M-commerce favorably.

Venkatesh and Davis [2000] extended the TAM to TAM2. User acceptance was found to be significantly influenced by social influence processes (subjective norm, voluntariness, and image) and cognitive instrumental processes (job relevance, output quality, result demonstrability, and perceived ease of use). TAM2 was empirically supported with longitudinal test data collected regarding four different systems at four organizations, two involving voluntary usage and two involving mandatory usage.

Venkatesh *et al.* [2003] proposed the unified theory of acceptance and use of technology (UTAUT) by reviewing eight models on user acceptance. The UTAUT aims to explain user intentions to use an information system and subsequent usage behavior. The theory holds that four key constructs (performance expectancy, effort expectancy, social influence, and facilitating conditions) are direct determinants of usage intention and behavior. Gender, age, experience, and voluntariness of use are posited to mediate the impact of the four key constructs on usage intention and behavior. The eight models reviewed are the TRA, the TAM, the motivational model, the theory of planned behavior, a model combining the TAM and the theory of planned behavior, the model of PC utilization, the innovation diffusion theory, and the social cognitive theory.

UTAUT was empirically tested and found to outperform the eight individual models.

3. A Model of Attitudes toward Mobile Phone Use in Public Places

As suggested by Venkatesh *et al.* [2003], social influence is one important determinant of usage intention and behavior. Therefore, the attitude of users about the social acceptance of public places and its use in public places will be important in affecting usage intention and behavior. In addition, demographic factors such as gender, age, experience, education, are posited to affect this attitude toward social use of mobile phones and usage behavior [Agarwal and Prasad, 1999; Porter and Donthu, 2006; Yang, 2005]. As indicated by the case of NTT Docomo i-mode [Kodama, 2002; Jonason *et al.*, 2001], which has been accepted in Japan more readily than in other countries [Hung and Yeh, 2007], we posit in our model that country is an important factor in affecting both the attitude of acceptance and usage behavior. Therefore, we posit country as a factor affecting attitude and usage, in addition to other demographic factors such as age, gender, experience, and education. Figure 1 shows the model proposed. In the model, the attitude of users is affected by users' demographic factors, specifically country, age, education, work status, and gender. In turn, the attitude of users affects users' usage of mobile phones.

3.1. *Attitudes about mobile phone use in public places*

The popularity of mobile phones has resulted in their increasing use in public places. People may communicate with one another through mobile phones anytime in public places such as restaurants, sidewalks, restrooms, and public transportation.

The use of mobile phones in public settings also caused the establishment of "simultaneous public and private spaces"

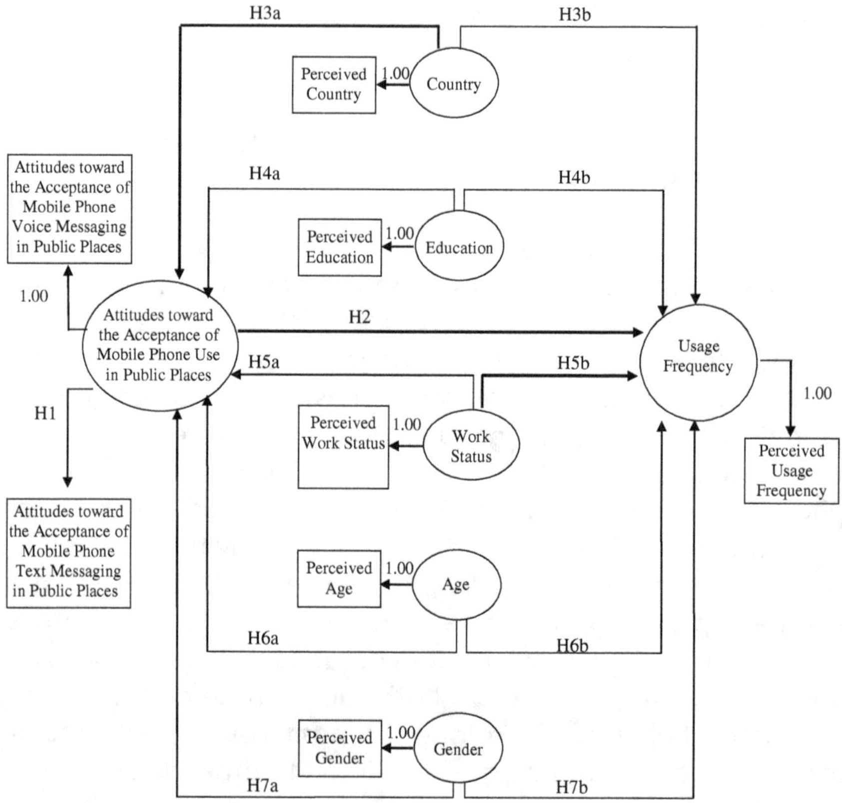

Figure 1. The hypothetical model of attitudes toward mobile phone use in public places.

[Ling, 2004]. Many individuals consider the mobile phone an extension of their physical selves [Campbell, 2007] and symbolic representatives of individual identities. According to Turner *et al.* [2007], mobile phone users differed in the extent to which they feel comfortable making and receiving calls in different social contexts, and are less annoyed by others using mobile phones in locations where they themselves feel most comfortable making calls. Researchers found that an individual in a social situation might have three types of responses to a mobile phone call [Weilenmann and Larsson, 2001]. The individual might move away from the social situation, remain in place but removed from the social engagement, or stay engaged socially. Murtagh [2001] noted that people try to create a private space in a limited public space when receiving mobile phone calls.

In many public places, people around the mobile phone users may be annoyed by the use of the mobile phones [Ling, 1997; Ling and Haddon, 2003]. Common complaints included disturbance and irritation of the quietness caused by the loud talk and the ringing of mobile phones in public places. Mobile phone users engaging in this behavior were regarded as discourteous, intrusive, and arrogant. In addition, people were concerned about the use of mobile phones in public places that might divert attention or caused danger. To limit improper uses, several solutions have been suggested, including imposing penalties, asking users to exercise self-discipline, posting notices in public places, providing structured guidelines on websites, and providing patrons with a list of areas in which use is permitted [Lever and Katz, 2007].

Most of the complaints about mobile phone use in social settings focused on the loud talking and ringing associated with the voice functions of mobile phones. Compared with voice messaging, text messaging offers a less intrusive form of communication in public places. Short message service (SMS) or text messaging with mobile phones has been become a preferred means of communication [Faulkner and Culwin, 2005]. The total figures for text messages sent in the United Kingdom in 2003 was 20.5 billion, and text messaging was viewed as a warm, personal, and cost-effective way to communicate. In France, for example, a study of mobile phone use among the population shows that 89% of the people under 40 send text messages, but only 47% of those over 40 send text messages [Observatoire Societal du telephone mobile, 2005]. Text messaging may have a different impact in social settings than voice communication because it is less intrusive than voice. Thus, both voice messaging and text messaging are important uses of the mobile phones.

The construct of "attitudes toward the acceptance of mobile phone use in public places" is a latent construct that will be observed through two different ways. We hypothesize the following:

H1: The construct "attitudes toward the acceptance of mobile phone use in public places" can manifest itself in two ways: (i) The degree to which an individual accepts mobile phone voice messaging in public places and (ii) The degree to which an individual accepts mobile phone text messaging in public places.

3.2. *The effect of attitude of mobile phone acceptance on mobile phone usage*

According to Ajzen and Fishbein [Davis *et al*., 1989], an individual's attitude toward the mobile phone usage is affected by his or her actual behavior. An individual's usage frequency of mobile phone use is affected by the individual's attitudes toward its social acceptance, or his or her attitude toward the acceptance of the use of mobile phones in public places.

Legris and Ingham [2003] reviewed studies conducted on the TAM, and indicated that several researchers were able to find a relationship between attitude and usage behavior. For example, in an empirical study of 230 users of an information technology innovation, Agarwal and Prasad [1999] confirmed that attitudes affected intentions to use and actual usage. Bajaj and Nidumolu [1998], on testing a longitudinal model with statistical panel data, found that a positive attitude toward using an information system led to increased usage of the information system. He also found that past usage affected ease of use, which in turn affected the attitude towards using an information system. Porter and Donthu [2006] tested the effect of age, education, income, and race on beliefs about the Internet, and attitudes toward and use of the Internet. They found that attitude toward internet usage was significantly and positively correlated with internet usage.

We propose the following hypothesis:

H2: Usage of mobile phones is affected by attitudes toward its use in public places. A user with a more positive attitude on accepting mobile phone use in public places will use the mobile phone more frequently.

3.3. *The effect of country on attitudes of acceptance and usage of mobile phones*

NTT Docomo i-mode [Kodama, 2002; Jonason *et al.*, 2001] was successful in Japan but its diffusion in other countries was slow [Hung and Yeh, 2007]. This suggests that country is a factor affecting attitude and usage. People in different countries differ in their social needs and also in their interpretations of appropriate social behavior. Ling and Haddon [2003] reported that mobile phone users in France, Germany, Italy, Spain, and the United Kingdom were least likely to leave their mobile phones switched on, when attending plays or shows. Licoppe and Heurtin [2001] showed that French people, in order not to be disturbed, would adopt different strategies to avoid giving their private mobile phone numbers to others.

Kleinman [2006] compared the use of mobile information and communication technologies in the cafés in France and the United States. Through participant observation and interviews, she found that cultural differences explained why mobile phones are not used as routinely in French cafés as they are in the United States. Individuals go to cafés in Paris to enjoy the ambience and the company of friends. It is common for people to sit at tables at sidewalk cafés, "watching the world go by". Although mobile phones are widely used in France, during Kleinman's participant observation in August 2004, only once was a person observed talking on a mobile phone inside of a café. The inside of the café was almost empty except for one table, and most customers were sitting outside. Kleinman noted that in the United States, by contrast, people talked on mobile phones in restaurants, churches, concerts, often times ignoring the person next to them.

As discussed in the introduction, according to census data, the number of mobile phone users was 69% of the US population at the end of 2005. In Europe, according to International Telecommunication Union [2005], the number of mobile phone users was 85.6 per 100 households in 2003, with 90.96 in Finland, 101.76 in Italy, 69.59 in France. People in different countries use mobile phones at different rates. Thus, we hypothesize the following:

H3a: Country affects attitudes toward the acceptance of mobile phone use in public places. Individuals in different countries have different attitudes toward the acceptance of mobile phone usage in public places.

H3b: Country affects the frequency of mobile phone usage. Individuals in different countries have different usage frequencies.

3.4. *The effect of education on attitudes of acceptance and usage of mobile phones*

Rice and Katz [2003] found that education had a significant influence on mobile phone and internet usage, and non-users and mobile phone users were more educated than internet users and users of both media. Hodge and Treiman [1968] found that one's education affects one's social participation patterns. Users' education level might affect the mobile phone usage level and the attitudes toward the acceptance of mobile phone use in public places. Robertson *et al.* [2007], using survey data collected in the United Kingdom, found that households with members who are better-educated and wealthier are less price-sensitive to internet services and more ready to adopt the services. Porter and Donthu [2006] found that older and less-educated individuals have lower perceived ease of use of the Internet, and a less positive attitude toward use of Internet. Dattee and Weil [2007], in analyzing the dynamics of social factors in technological substitutions, argued that innovators and early adopters "are always on top of their field's literature and with many connections and are therefore best reached by more direct communication media". According to Rogers [1995], early adopters were younger and more educated.

H4a: Education level affects attitudes toward the acceptance of mobile phone use in public places. Individuals with higher education are more likely to agree to the acceptance of mobile phone use in public places.

H4b: Education level affects mobile phone usage frequency. Individuals with higher education have higher mobile phone usage frequency.

3.5. *Effect of work status on attitudes of acceptance and usage of mobile phones*

Katz and Aspden [1998] found that ownership of mobile communication devices was determined by social location effects, such as the need to be in touch or being highly mobile, suggesting that working people might have a higher need for mobile phone use. In addition, Rice and Katz [2003] found that there was a significant correlation between work status and the use of mobile phones, and mobile phone users were more likely to work full-time. Busselle *et al.* [1999] looked at factors that affect usage of an innovation among faculty and staffs at Washington State University, and found that those owning more technologies were heavier users of the Internet. In addition, as working individuals might need to make and receive calls more frequently, they might also be more accepting toward the use of mobile phones in public places. We hypothesize the following:

H5a: Work status affects attitudes toward the acceptance of mobile phone use in public places. Individuals who are working are more likely to accept the use of mobile phones in public places.

H5b: Work status affects mobile phone usage frequency. Individuals who are working have higher mobile phone usage frequency.

3.6. *The effect of age and gender on attitudes of acceptance and usage of mobile phones*

Busselle *et al.* [1999] found that, in terms of age and gender, younger males were heavier users of the Internet. Rice and Katz [2003] found that mobile phone users were younger than non-users. Turner *et al.* [2007] found that younger respondents used their mobile phones more often, and older participants were less accepting toward the public use of mobile phones by others. Younger age groups have shown that they are more ready to adopt mobile phones as an important means of social communication [Aoki and Downes, 2003].

H6a: Age affects attitudes toward the acceptance of mobile phone use in public places. Younger individuals are more likely to agree to the acceptance of mobile phone use in public places.

H6b: Age affects the frequency of mobile phone usage. Younger individuals have higher mobile phone usage frequency.

From their research on the gendered nature of mobile phone culture in Israel, Lemish and Cohen [2005] concluded that men and women might have different attitudes toward mobile phones. Rakow and Navarro [1993] noted male and female users used the mobile phone differently. The mobile phone "seems to be an extension of the public world when used by men, an extension of the private world when used by women". Nysveen *et al.* [2005] found that social norms and intrinsic motives such as enjoyment were important determinants of intention to use among female users of mobile chat, whereas extrinsic motives such as usefulness and — somewhat surprisingly — expressiveness were key drivers among men.

Turner *et al.* [2007] found that female respondents were more accepting toward mobile phone use in public places, while male participants expressed greater annoyance with the public use of mobile phones. They also found that female respondents used their mobile phones more often than male respondents. DeBaillon and Rockwell [2005], however, did not find significant difference in the use of mobile phones between men and women among college and high school students.

We therefore hypothesize as follows:

H7a: Gender affects attitudes toward the acceptance of mobile phone use in public places. Males are less likely than females to agree to the acceptance of mobile phone use in public places.

H7b: Gender affects the frequency of mobile phone use in public places. Females have higher mobile phone usage frequency than males.

4. Methodology and Sample Characteristics

4.1. *Sample*

We conducted a survey among students in five countries: United States, France, Italy, Turkey, and Finland. The European countries were selected because they represent a cross-section of Europe politically, economically, and geographically, from Finland in the far North to Turkey in the South, with France and Italy in between. Comparisons among these countries and with the US were thought likely to be interesting. The survey was distributed to business students at universities in these countries. The survey included a statement that all respondents must either own or use a mobile phone for full-time, and so potential respondents who did not meet this criterion did not respond.

Table 2 shows the questions used to collect the demographic variables of the respondents. These variables include country, work status, gender, age, and education. In addition, for the country factor, the people administering the questionnaires noted the country in which the survey was conducted. The country factor was coded as 1 for United States, 2 for France, 3 for Italy, 4 for Turkey, and 5 for Finland.

A total of 1,079 questionnaires were collected, and on removing cases with incomplete data, 989 valid questionnaires were used in the analysis. There were 283 cases from USA (28.6%), 240 cases from France (24.3%), 194 cases from Italy (19.6%), 200 cases from Turkey (20.2%), 72 cases from Finland (7.3%). The characteristics of the sample revealed that more than half of the respondents were female (54.1%). Most of the respondents were between the ages of 20–25 (72.4%), with 7.9% under 20, 40.2% between 20 and 22, 32.2% between 23 and 25, 11.4% between 26 and 30, and 8.3% above 30. About 30% of the respondents were graduate students, while 70.3% were undergraduate students. About 45% of the respondents were working. About 34% had owned the mobile phone for more than five years, 39% had owned it for three to five years, 15% had owned it for two to three years, and 13% had owned it for less than two years. Table 2 also summarizes the demographic characteristics of the respondents in percentages.

In the survey (see Table 1), the following open-end questions were asked: Q1 — Where is using a mobile phone for voice inappropriate? Q2 — When is using a mobile phone for voice inappropriate? Q3 — Where is using a mobile phone for text messaging inappropriate? Q4 — When is using a mobile phone for text messaging inappropriate? Most respondents viewed it inappropriate to use voice messaging in movies, theaters, performance and concerts (592 cases),

Table 1. Items within the questionnaire for attitudes toward the acceptance of mobile phone use in public places.

Variable	Item	Question
Attitudes toward the acceptance of public mobile phone voice messaging in public places	Voice1	Cell phone voice calls on public transportation should not be prohibited
	Voice2	Cell phone voice calls while walking should not be prohibited
	Voice3	Cell phone voice calls in public restrooms should not be prohibited
	Voice4	Cell phone voice calls in restaurants should not be prohibited
Attitudes toward the acceptance of public mobile phone text messaging in public places	Text1	Cell phone for text messaging on public transportation should not be prohibited
	Text2	Cell phone for text messaging while walking should not be prohibited
	Text3	Cell phone for text messaging in public restrooms should not be prohibited
	Text4	Cell phone for text messaging in restaurants should not be prohibited

classroom (367 cases), quiet areas (194 cases), libraries (180 cases), cars (143 cases). Respondents noted that voice messaging should not be used when it is disturbing others (168 cases) or when it is inappropriate (72 cases), when talking to others (101 cases). On the other hand, fewer respondents view it inappropriate to use text messaging in cars (193 cases), classroom (121 cases), theatres, movies, concerts and performance (91 cases), libraries (11 cases), quiet areas (41 cases). They indicated text messaging should not be used during driving (201 cases), when it is dangerous (32 cases), and while talking with someone (80 cases). Answers to these questions show that most people agree to the acceptance of mobile phones when it is allowed legally, as in the case of driving in many countries, or when it is allowed socially, as in the case of inside restaurants.

Table 2. Demographic variables and respondent characteristics.

Variable	Questions	Coding	Percentages	
Work status	Are you currently working?	1 for yes, and 0 for no	Working	45.4
			Non-working	54.4
			NR	0.2
Gender	Are you male or female?	1 for male, 0 for female	Male	46.0
			Female	53.9
			NR	0.1
Age	How old are you?	Age in number of years	<20	7.7
			20–22	39.6
			23–25	31.2
			26–0	11.7
			>30	9.4
			NR	0.5
Education	Are you a graduate or undergraduate student?	1 for graduate, 0 for undergraduate	Graduate	29.7
			Undergraduate	70.3
Country		1 for US, 2 for France, 3 for Italy, 4 for Turkey, 5 for Finland	USA	28.6
			France	24.3
			Italy	19.6
			Turkey	20.2
			Finland	7.3

5. Results and Findings

The data were analyzed using LISREL [Jöreskog and Sörbom, 1989, 1999; Long, 1983], an instrument used for the analysis of structural equation models. LISREL is a useful tool in understanding the underlying dimension of a latent variable, such as the attitudes of the user toward the acceptance of mobile phone use in public places, and gives an accurate picture of the simultaneous effect of the demographic factors on the attitudes. In the structural equation model, a set of empirically operational indicators are defined for each dimension of each construct [Jöreskog, 1993]. Maximum likelihood estimation (MLE) was used to obtain the structural coefficients [Bollen and Long, 1993].

The final model is given in Figure 2. In the final model, the latent construct "Attitudes toward the Acceptance of Mobile Phone Use in Public Places" is made up of two significant indicators, "Attitudes toward the Acceptance of Mobile Phone for Voice Messaging in Public Places" and "Attitudes toward the Acceptance of Mobile Phone for Text Messaging in Public Places".

The adequacy of the model in Figure 2 is assessed using various measures [Kaplan, 2000; Loehlin, 1998; Diamantopoulos and Siguaw, 2000]. In testing structural equation models, the null hypothesis is set up as *a priori* not to be rejected, while the chi-square statistic tests whether the observed data fit the hypothesis of the proposed model, and a smaller chi-square value indicates a better fit. For small sample sizes that might have slightly departed from normality, instead of using the chi-squares, the chi-square per degree of freedom should be used. A ratio of approximately 5 shows a reasonable fit while a ratio between 1 and 2 is an excellent fit [Kaplan, 2000]. The ratio of the model in Figure 2 is 1.793 (chi-square = 10.76 with 6 degrees of freedom), indicating a very good fit.

Other measures of fit include the goodness of fit index (GFI) and normed fit index (NFI). Both the GFI and NFI are always

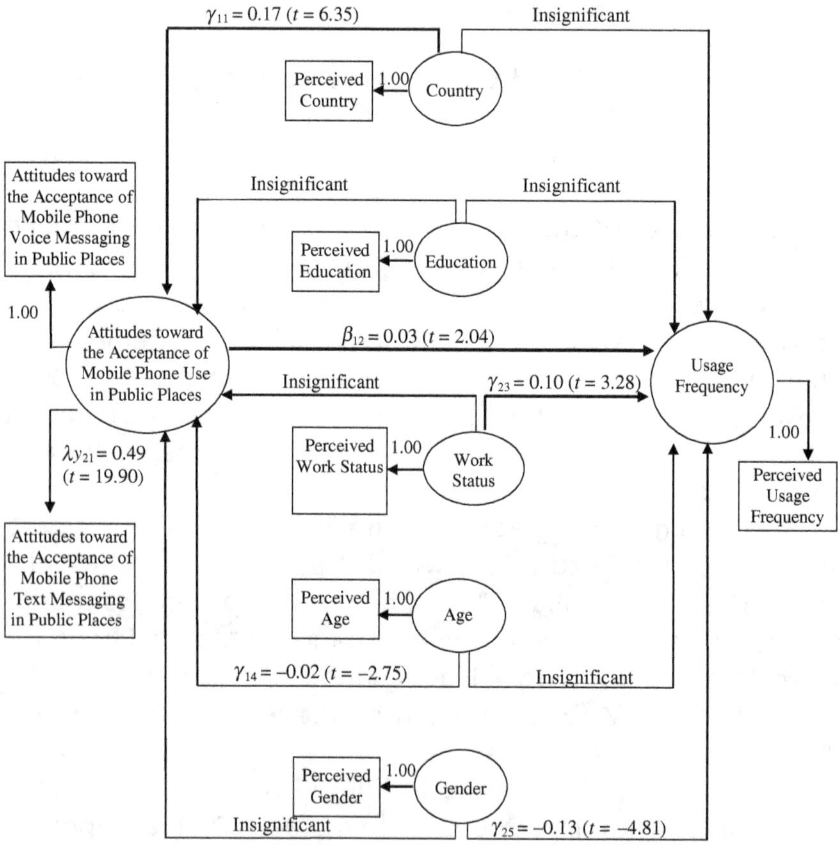

Figure 2. Final estimates of the model of attitudes toward mobile phone use in public places.

between 0 and 1, with 1 indicating a perfect fit while any value above 0.9 suggesting a good fit [Bentler and Bonett, 1990]. The model has a GFI of 1.00 and a NFI of 0.99. This shows a good fit. The adjusted goodness of fit (AGFI) is 0.98. This again shows a good fit. Similarly, the non-normed fit index (NNFI) and the comparative fit index (CFI) are two additional measures ranging from 0 to 1, where values greater than 0.9 represent a good model fit. The NNFI and CFI for the model are 0.97 and 0.99, respectively. Finally, the structural equation model was assessed using the root mean error of approximation (RMSEA) and the standardized root mean square residual (SRMR). The RMSEA is a measure of the discrepancy per degree of freedom for the model. A RMSEA value of 0.05 or less indicates a good model fit [Browne and Cudeck, 1993, p.144]. The SRMR is an absolute fix index sensitive to misspecification, and a value of 0.05 or below indicates a good fit [Hu and Bentler, 1999]. The model has a RMSEA of 0.028 and a SRMR of 0.015, thus indicating a good model fit. Thus, the GFI, AGFI, NFI, NNFI, CFI, SRMR, and RMSEA all indicate that the model has a good fit.

Figure 2 summarizes the maximum likelihood parameter estimates and *t*-values for the model constructs shown in Figure 2. For the latent variables "Attitudes toward the Acceptance of Public Mobile Phone Voice Messaging in Public Places", λy_{11} is set to 1 to define the unit of measurement for "Attitudes toward the Acceptance of Mobile Phone Use in Public Places". As indicated, λy_{21} ($t = 19.90$) is 0.49 and is significant at the 0.05 level. This suggests that "Attitudes toward the Acceptance of Public Mobile Phone Voice Messaging in Public Places", together with "Attitudes toward the Acceptance of Public Mobile Phone Text Messaging in Public Places", are observable measures of the latent variable "Attitudes toward the Acceptance of Mobile Phone Use in Public Places". Thus, H1 is accepted. Since the coefficients λy_{21} is positive, this shows that if a person agrees to the acceptance of mobile phone text messaging in public places, the person would also agree to the acceptance of mobile phone voice messaging in public places. As the magnitude of the coefficient λy_{21} is less than 1, this shows that "Attitudes toward the Acceptance of Public Mobile Phone Text Messaging in Public Places" is a less important indicator than "Attitudes toward the Acceptance of Public Mobile Phone Voice Messaging in Public Places" in measuring "Attitudes toward the Acceptance of Mobile Phone Use in Public Places".

Figure 2 also shows the factors affecting the attitudes toward the acceptance of mobile phone use in public places. As shown in Figure 2, β_{12} is 0.03 ($t = 2.04$) and is significant at the

0.05 level, showing that usage frequency is positively affected by the attitudes toward the acceptance of mobile phone use in public places. This indicates that individuals who are more likely to agree to the acceptance of mobile phone use in public places tend to have higher mobile phones usage frequency. Thus, H2 is supported.

In addition, γ_{11} is 0.17 ($t = 6.35$) and is significant at the 0.05 level, showing that respondents in different countries differ in their attitudes toward the acceptance of mobile phone use in public places. Thus, H3a is supported. In addition, γ_{14} is −0.02 ($t = -2.75$) and is significant at the 0.05 level, showing that age negatively affects the attitude toward the acceptance of mobile phone use in public places. Younger respondents were more likely to agree to the acceptance of mobile phone use in public places. Thus, H6a is supported. The effects of education, work status, and gender, however, on attitudes toward the acceptance of mobile phone use in public places are insignificant. Thus H4a, H5a, and H7a are not supported.

In Figure 2, γ_{23} is 0.10 ($t = 3.28$) and is significant at the 0.05 level, showing that work status positively affects the usage frequency of mobile phones. Respondents who were working had higher usage frequency. Thus, H5b is supported. Further, γ_{25} is −0.013 ($t = -4.81$) and is significant at the 0.05 level, showing that females (gender is coded as 1 for male, and 0 for females) had higher mobile phone usage frequency. Thus, H7b is supported. However, the effects of country, education, and age on mobile phone usage frequency are insignificant. Thus, H3b, H4b, and H6b are not supported.

A chi-square statistic test was further performed to see if the insignificant relationships should be removed to improve the existing model. On removing the insignificant relationships in the original full model, the revised model has a RMSEA of 0.03. The chi-square difference test was performed to compare the original full model with the revised model. The chi-square of the full model is 10.76 with 6 degrees of freedom while the chi-square of the revised model is 23.01

with 12 degrees of freedom. Applying the likelihood ratio test for the comparison of models as suggested by Bollen [1989, p. 292], we found the likelihood test statistic is 12.25 (= 23.01 – 10.76) with 6 degrees of freedom (= 12 – 6). This chi-square estimate is statistically insignificant at the 0.05 level (the critical value of chi-square with 6 degrees of freedom at significance level of 0.05 is 12.59), indicating the revised model is not a better model than the full model shown in Figure 2.

6. Implications of Our Research Findings

In this study, as consistent with theories on attitudes and usage behavior (e.g., the Fishbein model and the TAM [Davis *et al.*, 1989]), we have found that usage frequency is affected positively by the attitudes toward the acceptance of mobile phone use in public places. Thus, it is important for mobile phone providers to foster positive attitudes of users toward acceptance of mobile phones in public places in order to increase users' usage.

In turn, the attitudes toward the acceptance of mobile phone use in public places are affected by country and age. Attitudes toward acceptance differ among countries. For mobile phone providers, analysis is required to identify the culture in each country in order to develop the appropriate advertising and promotion strategies as well as mobile phone plans that are targeted to the needs of the users in that country. For example, Turkey indicated the highest attitude toward the acceptance of mobile phone use in restaurants and the lowest toward its acceptance on public transportation. That suggests that, in Turkey, providers might launch mobile commerce functions such that advertisers might launch ads that are more affiliated with user settings in restaurants than on public transportations.

This study shows that younger respondents are more likely to agree to the acceptance of mobile phone use in public places. This implies mobile phone developers can target at young people and develop mobile phone functions that will meet the needs of this target group. Examples include social networking functions and mashup functions, or web-based application that combines content and functionality from a variety of sources using different technologies (http://www.tamingthebeast.net/articles6/web2-mashups-social-network.htm).

In addition to attitudes, usage was found to be affected directly by gender. Females tend to use mobile phones more frequently. This might suggest that the needs of the female customers might be different from those of the male customers, and a mobile phone plan with more minutes might be developed for the female market. Usage was also found to be affected by work status. Working people tend to use mobile phones more frequently. Providers may develop plans with more minutes to appeal to the working individuals.

7. Limitations and Future Research

One limitation of our research is we only include variables such as demographic factors, attitudes of acceptance of mobile phone usage in public places, and usage frequency. Future research might include additional variables such as perceived ease of use, perceived usefulness, as well as tendency to use mobile phones, in order to understand the model relationships among these elements for mobile phones.

A second limitation of this study is our sample consists of college students, with more than 90% between the ages of 18 to 30, 8% between the ages of 31 to 40, and 2% over 40. Future research might conduct the survey on a larger group of respondents, with a broader age range, so that the effect of age differences may be better observed. Future research might also conduct studies among working professionals with different levels of education levels, income, work position in order to better assess the effects of these variables. A third limitation is our survey is primarily focusing on users. Future research might include non-users as well as users, to understand how both groups assess mobile phone usage in public, and the factors affecting their attitudes of acceptance.

Based on our results, compared with the respondents in United States, respondents in other countries such as Turkey, Italy, and Finland were more likely to agree to the acceptance of mobile phone use in restaurants. More detailed study would need to be conducted to investigate the differences in cultures among these countries in order to understand the difference in attitudes toward mobile phone use among these countries. Future research might address the analysis of the usage behavior such as the time of the call, and usage duration, and their relationships to the time of the call and the location of the call.

8. Conclusion

This paper serves as a basis for enhanced understanding of acceptance and diffusion of new technologies. Several models and theories about new technology use already exist, including the TAM [Davis *et al.*, 1989] and diffusion of innovations theory [Rogers, 1995]. The results of this paper indicate that other factors might be considered when examining the use of new technologies. For example, are attitudes about the use of technology in different social settings an inhibitor to the acceptance of new technology? Does the diffusion of new technology in different countries depend on the country as well as these attitudes? Exploring these and similar questions could yield interesting results in the complex problem of technology acceptance and diffusion.

In this study, we have developed a structural equation model that helps us better understand the simultaneous relationships among users' demographic factors, mobile phone usage frequency, and attitudes toward mobile phone use. Our results have shown that usage depends on attitudes toward the acceptance of mobile phone use in public places, gender, and work status while attitudes toward the acceptance of mobile phone use in public places differ across country and age groups. Future research might address the increasing use of mobile phones for various purposes across different countries, and how mobile phone attitudes and usage behavior might be related to others factors such as self-esteem, culture, values. This research will help us understand how mobile phones might be better used as a means of communication and information exchange in our society.

References

Agarwal, R. and Prasad, J. (1999). Are individual differences germane to the acceptance of new information technologies. *Decision Sciences*, **30**(2):361–391.

Aoki, K. and Downes, E. J. (2003). An analysis of young people's use of and attitudes toward cell phones. *Telematics and Informatics*, **20**: 349–364.

Bentler, P. M. and Bonett, D. G. (1990). Comparative fit indices in structural models. *Psychological Bulletin*, **107**(2):238–246.

Bollen, K. A. (1989). *Structural Equation with Latent Variables*. Wiley and Sons, NY.

Bollen, K. A. and Long, J. S. (1993). *Testing Structural Equation Models*. Sage Publications, Newbury Park.

Bradley, S. and Sandoval, M. (2002). Case study: NTT DoCoMo: The future of the wireless Internet? *Journal of Interactive Marketing*, **16**(2):80–96.

Browne, M. W. and Cudeck, R. (1993). Alternative ways of assessing model fit. *Testing Structural Equation Models*, Eds. K. A. Bollen, and S. J. Long. Sage Publications, Newbury Park, pp. 136–162.

Busselle, R., Reagan, J., Pinkleton, B. and Jackson, K. (1999). Factors affecting internet use in a saturated-access population. *Telematics and Informatics*, **16**(1/2):45–58.

Campbell, S. (2007). Perceptions of mobile phone use in public settings: A cross-cultural comparison. *International Journal of Communication*, **1**:738–757.

Central Intelligence Agency (2007). The CIA World Factbook 2008, New York, Skyhorse Publishing.

Choi, H., Kim, J., Lee, M., Giaglis, G., Naruse K., Constantiou, I. D., Fife, E. and Mayhew, G. (2007). Contribution to quality of life: Cross-national validation of new metrics for mobile data service technology in Korea and Japan. *International Journal of Innovation and Technology Management*, **4**(2):191–208.

Dattee, B. and Weil, H. B. (2007). Dynamics of social factors in technology substitutions. *Technology Forecasting & Social Change*, **74**(5): 579–607.

Davis, F. D., Bagozzi, R. P. and Warshaw, P. R. (1989). User acceptance of computer technology: A comparison of two theoretical models. *Management Science*, **35**(8):982–1003.

DeBaillon, L. and Rockwell, P. (2005). Gender and student-status differences in cellular telephone use. *International Journal of Mobile Communications*, **3**(1):82–98.

Diamantopoulos, A. and Siguaw, J. A. (2000). *Introducing LISREL: A Guide for the Uninitiated*. Sage Publications, London.

Faulkner, X. and Culwin, F., (2005). When fingers do the talking: A study of text messaging. *Interacting with Computers*, **17**: 167–185.

Green, T. (2007). Mobile subscriber base to pass 3bn in weeks. *Mobile Entertainment*, June 27. Available at http://www.mobile-ent.biz/news/27780/Mobile-subscribers-to-hit-3bn-in-weeks.

Hacklin, F., Raurich, V. and Marxt, C. (2005). Implications of technological convergence on innovation trajectories: The case of ICT industry. *International Journal of Innovation and Technology Management*, **2**(3):313–330.

Hodge, R. W. and Treiman, D. J. (1968). Social participation and social status. *American Sociological Review*, **33**:723–740.

Hu, L. and Bentler, P. M. (1999). Cutoff criteria for fit indexes in covariance structure analysis: Conventional criteria versus new alternatives. *Structural Equation Modeling*, **6**:1–55.

Hung, C. L. and Yeh, S. Y. (2007). The diffusion of I-mode overseas: A perspective of institutional differences. *International Journal of Innovation and Technology Management*, **4**(1):69–86.

International Telecommunication Union (2005). Europe and CIS's telecommunication/ICT markets and trends, 2003/2004. Available at http://www.itu.int/ITU-D/ict/statistics/, pp. 1–14.

Jackson, C. M., Chow, S. and Leitch, R. A. (1997). Toward an understanding of the behavioural intention to use an information system. *Decision Sciences*, **28**(2):357–389.

Jonason, A. and Eliasson, G. (2001). Mobile internet revenues: An empirical study of the I-mode portal. *Internet Research: Electronic Networking Applications and Policy*, **11**(4):341–348.

Jöreskog, K. G. (1993). Testing structural equation models. *Testing Structural Equation Models*, Eds. K. Bollen and J. S. Long, Sage Publications, Newbury Park.

Jöreskog, K. G. and Sörbom, D. (1989). *LISREL 7: A Guide to the Program and Applications*, 2nd edn., SPSS Inc., Chicago, .

Jöreskog, K. G. and Sörbom, D. (1999). *Prelis 2: User's Reference Guide*, Scientific Software International Inc., Lincolnwood, IL.

Kaplan, D. (2000). *Structural Equation Modeling: Foundations and Extensions.* Sage Publications, London.

Katz, J. E. and Aspden, P. (1998). Theories, data, and potential impacts of mobile communications: A longitudinal analysis of U.S. national surveys. *Technological Forecasting and Social Change*, **57**:133–156.

Kim, M. S. (2005). Household-use versus business-use demand diffusion of telephone and Internet access service in Korea. *International Journal of Innovation and Technology Management*, **2**(1):57–75.

Kim, S. H. (2008). Moderating effects of job relevance and experience on mobile wireless technology acceptance: Adoption of a smartphone by individuals. *Information & Management*, **45**(6):387–393.

Kodama, M. (2002). Transforming an old economy company into a new economy success: The case of NTT DoCoMo. *Leadership & Organization Development Journal*, **23**(1):26–39.

Kleinman, S. (2006). Café culture in France and the United States: A comparative ethnographic study of the use of mobile information and communication strategies. *Atlantic Journal of Communication*, **14**(4):191–210.

Legris, P. and Ingham, J. (2003). Why do people use information technology? A critical review of the Technology Acceptance Model. *Information & Management*, **40**(3):191–204.

Lemish, D. and Cohen, A. A. (2005). On the gendered nature of mobile phone culture in Israel. *Sex Roles*, **52**(7/8):511–521.

Lever, K. M. and Katz, J. E. (2007). Cell phones in campus libraries: An analysis of policy responses to an invasive mobile technology. *Information Processing and Management*, **43**:1133–1139.

Licoppe, C. and Heurtin, P. (2001). Managing one's availability to telephone communication through mobile phones: A French case study of the development dynamics of mobile phone use. *Personal and Ubiquitous Computing*, **5**:99–108.

Ling, R. (1997). One can talk about common manners! The use of mobile telephone in inappropriate situations. *Themes in Mobile Telephony: Final Report of the COST 248 Home and Work Group*, Ed. L. Haddon. Telia, Stockholm.

Ling, R. (2004). *The Mobile Connection: The Cell Phone's Impact on Society (Interactive Technologies)*, 3rd edn. Elsevier, Morgan Kaufmann Publishing, San Francisco.

Ling, R. and Haddon, H. (2003). Mobile telephony, mobility and the coordination of every day life. *Machine That Become Us: The Social Context of Communication Technology*, Ed. J. Katz. Transaction Publishers, New Brunswick, NJ, pp. 245–266.

Loehlin, J. C. (1998). *Latent Variable Models: An Introduction to Factor, Path, and Structural Analysis*, 3rd edn. Lawrence Erlbaum, London.

Long, J. S. (1983). *Covariance Structure Models: An Introduction to LISREL*, Sage Publications, Newbury Park, CA.

Moedritscher, G. and Mussnig, W. (2005). Evaluating the long-term benefit of CRM-systems: A cost-oriented approach. *International Journal of Innovation and Technology Management*, **2**(4):367–389.

Murtagh, G. M. (2001). Seeing the "rules": Preliminary observations of action, interaction and mobile phone use, *Wireless World: Social and Interactive Aspects of the Mobile Age*. Eds. B. Brown, R. Harper and N. Green, Springer-Verlag, London, pp. 81–91.

NTT DoCoMo (2008). I-mode, http://www.nttDoCoMo.com/services/imode/index.html.

Nysveen, H., Pedersen, P. E. and Thorbjørnsen, H. (2005). Explaining intention to use mobile chat services: Moderating effects of gender. *Journal of Consumer Marketing*, **22**(5):247–256.

Observatoire Societal du telephone mobile (2005). AFOM, SOFRES TNS, Novembre. Available at http://www.tns-sofres.com/etudes/pol/031105_telmobiles.htm.

Porter, C. E. and Donthu, N. (2006). Using the Technology Acceptance Model to explain how attitudes determine Internet usage: The role of perceived access barriers and demographics. *Journal of Business Research*, **59**(9):999–1007.

Rakow, L. and Navarro, V. (1993). Remote mothering and the parallel shift: Women meet the mobileular phone. *Critical Studies in Mass Communication*, **20**(3):144–157.

Ratten, V. and Ratten, H. (2007). Technological Innovations and m-Commerce Applications. *International Journal of Innovation and Technology Management*, **4**(13):1–14.

Rice, R. E. and Katz, J. E. (2003). Comparing internet and mobile phone usage: Digital divides of usage, adoption, and dropouts. *Telecommunications Policy*, **27**:597–623.

Rogers, E. M. (1962). *Diffusion of Innovations*. Free Press, Glencoe.

Rogers, E. M. (1995). *Diffusion of Innovations*, 4th edn., Free Press, NY.

Shim, J. P. (2005). Korea's lead in mobile cellular and DMB phone services, *Communications of the AIS*, **15**:555–566.

Turner, M., Love S. and Howell M. (2007). Understanding emotions experienced when using a mobile phone in public: The social usability of mobile (cellular) phones. *Telematics and Informatics,* **25**(3):201–215.

Venkatesh, V. and Morris, M. G. (2000). Why don't men ever stop to ask for directions? gender, social influence, and their role in technology acceptance and usage behavior. *MIS Quarterly*, **24**(1): 115–139.

Venkatesh, V. and Davis, F. D. (2000). A theoretical extension of the Technology Acceptance Model: Four longitudinal field studies. *Management Science*, **46**(2):186–204.

Venkatesh, V., Morris, M. G., Davis, G. B. and Davis, F. D. (2003). User acceptance of information technology: Toward a unified view. *MIS Quarterly*, **27**(3):425–478.

Wallace, B. (2006). 30 countries passed 100% mobile phone penetration in Q1. Telecommunications Online (http://www.telecommagazine.com), June 9.

Weilenmann, A. L. and Larsson, C. (2001). Local use and sharing of mobile phones. *Wireless World: Social and Interactive Aspects of the Mobile Age*, Eds. B. Brown, R. Harper, N. Green, B. Brown, R. Harper, N. Green. Springer-Verlag, London, pp. 99–115.

Chapter 12

Emerging Applications: Cyberbullying

Orawit Thinnukool, Pattaraporn Khuwuthyakorn, and Ruthaychonnee Sittichai

1. Introduction

In recent years, a considerable amount of research has been published and has drawn our attention to the fact that advances in ICTs, including smartphones and other mobile devices together with advanced high-speed internet connectivity, have changed societies around the world. Adolescents have trended to adopt these technologies from an early age, resulting in cyberspace becoming thoroughly embedded in their lives, especially social network sites (SNS).

Over the last decade, the times spent online, collected from ASEAN countries and those from western countries, have been similar. Many adolescents spend more time online, either for entertainment, education, communication, or social interaction. These activities may expose them to cyberbullying. This circumstance facilitates opportunities for adolescents to experience cyberbullying and its negative impacts.

Cyberbullying is an aggressive behavior with intent to harm people that takes place over the Internet [Smith *et al.*, 2008; Huang and Chu, 2010]. It includes sending or posting harmful anonymous manuscripts without author identity. It includes sending or posting harmful or engaging in other forms of social aggressive using the Internet [Willard, 2005]. Sharing personal information of someone such as photographs or videos is all associated with cyberbullying [Mishna *et al*., 2012]. Several researchers have defined cyberbullying as behavior that is associated with bullying through e-mail, short messaging via internet sites, or through a text message sent to a cell phone or any electronic device [Agatston *et al*., 2007; Li, 2008].

Many researchers are concerned about this problem and have reported on cyberbullying via several channels. Patchin and Hinduja [2006] reported that cyberbullying frequently occurs to adolescents, especially via SNS, such as Facebook, Twitter, Instagram, and other platforms. These products are tools that probably cause cyberbullying [Lenhart *et al*., 2011; Cassidy *et al*., 2013; Elizabeth and Robin, 2015]. Corresponding to the study of Charles *et al*. [2013], the major cause of engagement for cyberbullying of adolescents was the use of SNS.

Kwan and Skoric [2013] found the relationship between the use of social networks like Facebook and the number of cyber victims. Other studies found that cyberbullying occurs via the use of other internet-connected devices such as smartphones [Bianchi and Phillips, 2005; Ehrenberg *et al*., 2008]. So, these evidences have shown that cyberbullying covers all of the behavior that is associated with bullying via the Internet, and the probability of cyberbullying occurrences has also been claimed to involve various issues. It is likely that offenders who participate in cyberbullying also exhibit harmful behavior such as deception or defamation in the real world. Exceptionally, 84% of cyber victims felt angry when they knew their offenders [Wolak *et al*., 2006; Hinduja and Patchin, 2010].

A report about the relationship between cyber victimization and age states the highest number of offenses is at ages between 13 to 15 years old [Tokunaga, 2010]. It has been found that boys tended to bully more often than girls and it tended to occur in various contexts [Noam and Michal, 2015].

Regarding Southeast Asian countries, the report of Sittichai and Smith [2015] shows that "gender differences are mostly in line with western findings, namely that boys are more often perpetrators of bullying whereas there is much less of a gender difference in being a victim of bullying".

In Thailand, more than half of the victims (56%) reported the bully to their friends, while only 37% of the victims reported the bully to their parents, teachers (32%), and siblings (29%) [Tapanya, 2006]. Whereas, the report of Laeheem *et al.* [2008] shows that one-third of surveyed students admitted that they had bullied other children. There is a very interesting result of this research: the classroom was the most frequent location to bully.

Many researchers, nowadays, are still working on the tool for anti-cyberbullying, or on a tool to educate or to automatically detect cyberbullying. Several researchers had developed systems such as a game to educate players about cyberbullying, raising awareness, and understanding among the users. The results show that such the game successfully encouraged users to learn and raise awareness about cyberbullying.

Recently, there was a study that brought up an idea to develop a system for automatically detecting cyberbullying. The system would automatically monitor cyberbullying over social networking sites [Royen *et al.*, 2015]. The study surveyed many experts for their opinions and what functionality and desirability would be required in the system. Although this study did not produce any real system, the conceptual idea provides a good suggestion for future development. Their conceptual framework includes several functions such as the system should have a function for contacting victims and helping victims and the system should provide suitable resources for youth guidance.

Additionally, there are many research studies focused on the development tools for cyberbullying. Nadia *et al.* [2015] applied a multimedia application which was able to run on a personal computer. The result of this study claims that the perceived cyberbullying knowledge of adolescents after watching multimedia has increased.

Moreover, GroupCast LLC [2015] used the cyberbullying hotline, which is a mobile application that helps users to protect themselves against bullying. One useful function is the function to monitor students when they need consultation on cyberbullying: the application is a system for communication such as messaging or sending phone numbers like Line. As a result, students who use the application have a channel to consult with experts.

Furthermore, there are several programs which provide tools for consulting. No-Trap, Web Quest, and Online "Pest-koppenstoppen" are some examples of these programs, which provide intervention for adolescents in terms of helping them answer on cyberbullying questions [Menesini *et al.*, 2015; Brewer, 2011; Jacobs *et al.*, 2014]. The outcome is successful when adolescents had tools that could help them to answer their cyberbullying questions like an expert. These programs are useful tools for educating adolescents to understand and encourage them to be aware of the harm of cyberbullying. An up-to-date mobile application, STOPit, is another example of a program that presents the idea of sharing cyberbullying phenomena in each area. Taking a photo, recording sound, and chatting are embedded functions that help users to protect themselves from cyberbullying [STOPit cyberbully, 2015].

However, these days, only a few observed researches have been conducted involving mobile applications as a tool to enlighten users on cyberbullying knowledge. There are 38 CBMAs available in Apple's App Store. Most of them are used for educating users about the content or literacy of cyberbullying, and some of them have a suggestion function for the cyberbullying victims. These mobile applications are developed in many countries, but none of them are from the Thailand [Online Apple Store, 2016].

In Thailand, there was an observation regarding tools for anti-cyberbullying or monitoring cyberbullying. Some issues were found to involve websites for providing cyberbullying contents and some involved TV shows, such as the Thai government channel. These channels broadcast cyberbullying contents for teaching Thai adolescents to beware of cyberbullying in daily life. The question is why Thai people do not use tools or applications from outside of the country; the language barrier is the major problem of Thai people since the majority of Thais do not communicate in English. So, Thai adolescents do not fully understand the language. As a result, when considering tools for this purpose, nowadays, Thailand still needs to develop tools in Thai for monitoring or preventing cyberbullying.

So, the question is: should a mobile cyberbullying application that can provide the useful functions as mentioned earlier be developed in the Thai social context. Mobile devices may be suitable for providing information on cyberbullying for Thai adolescents, as mobile devices are very popular in Thailand.

The research stems from the concept of the development of a mobile application called CBMA which aims to educate and provide cyberbullying contents to adolescents. The estimated outcomes from this research is the CBMA can promote a better understanding of cyberbullying to Thai adolescents, while Thai adolescents would have a mobile-based tool to educate and encourage themselves against cyberbullying. Thus, in this study, the objectives are as follows:

(1) To investigate Thai adolescents' perceived knowledge of cyberbullying.
(2) To develop the CBMA based on iOS operating system.
(3) To evaluate users' feedback regarding GUI, system performance, confidence, and attitude on the CBMA.

Because the development proposal of the CBMA needs to support both high school students and undergraduate students, it is important to make sure that the overall user feedback regarding GUI, system performance, confidence, and attitude on the CBMA between two groups are not different. In this paper, questionnaire outcomes were analyzed by using independent sample *t*-test analysis on *R* program. Finally, the following hypotheses were examined:

H0: The average of agreement level of GUI preference, system performance, confidence, and attitude on the CBMA between two groups are not different.

H1: The average of agreement level of GUI preference, system performance, confidence, and attitude on the CBMA between two groups are different.

2. Materials and Methods

Types of Bullying Behavior. Bullying behavior is the beginning of harmful behavior that may occur either via cyber sites or SNS, and can lead to bullying behavior in the real world. Corresponding to this, the paper of Kowalski *et al.* [2016] says "traditional bullying, a student with disabilities is at particular risk for cyberbullying victimization".

Many physiologists describe "bullying" as an aggressive behavior that occurs when somebody is exposed to negative actions on others. Physiologists separate types of bullying that can be experienced by many groups such as children, adolescents, and adults. Different types of bullying defined by physiologists are similar to the ways in which they occur. The behavior of bullying includes physical bullying, verbal bullying, social bullying, and cyberbullying [National Centre Against Bullying, 2016].

Thus, in this research, the information on cyberbullying was gathered and separated into four categories presented by cyberbullying behaviors and the corresponding characteristics of each action, as shown in Table 1. Note that the information on each group of cyberbullying behavior in Table 1 corresponds to the CBMA.

Table 1. Cyberbullying groups which are separated by behavior.

Cyberbullying group	Cyberbullying behavior	Characteristic/Definition
1. Defamation, disparagement	Lying	Speaking or telling any false things, or other
	Imputation, aspersion, slander	Imputation, aspersion, or slandering on electronic media that causes another to be damaged
	Profanity	Speaking disrespectful or negative notation
	Revile, abuse	Being rude that causes another to be annoyed, doubtful, or mentally distressed
	Innuendo, sarcasm, satire	Mocking and taunting that make another uncomfortable
	Prate prattle	Speaking unintelligibly, unreliably, or telling unfounded information
	Posting messages that are not true which dishonors others	Posting information that causes mental damage or leads to physical damage
	Distribution information that causes others suffer profanity or disdain	Providing information with no evidence or facts that causes another prone to damage, or suffering, or hatred

(*Continued*)

Table 1. (*Continued*)

Cyberbullying group	Cyberbullying behavior	Characteristic / Definition
2. Impersonating someone else in a negative way	Stealing passwords	Steeling passwords of others by guessing or using tools in cyberspace
	Information filtering	Using tools in cyberspace to filter passwords or personal information of others without permission
	Creating fake accounts to be another	Impersonating someone by using his/her password to commit crimes in various fields and bring damage to the password owner
	Creating situation that encompasses people to hate each other	Creating either a true or not true situation that encompasses people to hate each other using a cyber way
3. Revealing of personal information of another without permission	Information hacking	Taking important information or passwords of another without permission
	Imitating someone else's profile	Imitating someone else's profile to commit crimes that causes negative impact and damages

Disclosure of confidential information	Publishing important information of another on cyber media without permission of the information owner
Editing information or images of another	Publishing personal information or pictures of another without permission that causes damages or humiliation
Blocking others from seeing the actual information	Disclosure or revealing of true information mixed with falsehood to make others misunderstand the whole truth
Deleting, or blocking, or destruction of facts data	Offenders intend to destroy information that are believed to be evidence of any crime on the network or computer system
Blocking someone to access his/her profile	Offenders run away by blocking others to get access to the source of information (such as Facebook, Line)

(*Continued*)

Table 1. (*Continued*)

Cyberbullying group	Cyberbullying behavior	Characteristic / Definition
4. Aspects related to sexuality	Posting sexual abuse messages to others	Posting words related to sexual harassment which causes suffering or discomfort to another
	Having text or sound of sexual **harassment**	Send offending text or inappropriate **video clips to victims without consent e.g., Sexphone**
	Receiving pornography	Victims receive pornography which is harassment from offenders
	Posting sexually related pictures that damage others	Posting pornographies which are edited and cause damage to others
	Posting pictures with someone to provoke the anger of another	Posting sexually related pictures to cause a breakage between individuals
	Youth-abusive sexual acts procured by fraud or false pretenses	Violation of the rights of others in related physical and emotional sexuality in different ways

3. Cyberbullying Mobile Application (CBMA)

The CBMA has been developed on the iOS operating system platform because majority of Thai adolescents prefer using iOS rather than Android. To develop the mobile application, the development followed the conceptual framework [GroupCast LLC, 2015], used together with useful information on cyberbullying from the study of Royen *et al.* [2015]. The study indicated some parts of the system which included important functions to help victims, provide suitable resources and consultation, and be a source of information to contact experts by making direct phone calls or messaging from the application.

Since its development, the CBMA has followed great suggestions from users; the CBMA is supportive when users have encountered a bullying problem, and also cyberbullying content provided by the application is beneficial for users. Moreover, sharing infographics related to each cyberbullying technique on social sites may notify users to spread the knowledge of cyberbullying to others.

In this version of the CBMA, no special function was released to Thai adolescents because it is the first version in development. Table 2 shows each function of the CBMA to provide more understanding of the application's functionalities.

After completing the final version with software testing, the CBMA was uploaded into Apple's App Store. The screen shots in Figure 1 indicate the screens and functions of the application. Figure 1(a) shows the title in the page of the application, Figure 1(b) shows a screenshot for login (as a guest or login via Facebook) to identify the user, whereas Figure 1(c) is the main page which separates cyberbullying into four main groups corresponding to cyberbullying behaviors in Table 1.

The screenshot in Figure 1(d) presents a list of cyberbullying behavior to provide users keywords for their search information, while Figure 1(e) indicates cyberbullying information such as cyberbullying behavioral characteristics, definition. The screenshot in Figure 1(f) shows the cyberbullying infographics

Table 2. Each function of the CBMA according to the functional requirement.

Function	Propose	How to use
(a) Sharing cyberbully information	User can immediately search by keyword to see cyberbully information	User has to go to searcher function. After seeing the cyberbully information, user clicks to share the detail via Facebook, message, and note
(b) Calling to consult aid agencies	User can call consultant from an organization in Thailand which provides hotline.	When user reads the cyberbully information on each behavior, user can immediately call the hotline
(c) Taking a photo	User can take a photo when encountering cyberbullying, when the photo can be evidence of the act	Go to camera function and take a photo (the photo has been recorded into hidden memory in the application)
(d) Setting	User can setup content privacy which needs authentication to login via Facebook	Go to Setting, and user can set the privacy of authentication which connects to Facebook

Figure 1. Screen shots of graphical user interface demonstrate functionalities.

via Facebook, when users want to alert others about cyberbullying. Furthermore, the purposes of sharing are to inform about the legal punishment and notify warnings when users encounter cyberbullying: they will share each infographic item on SNS to notify the cyberbully.

4. Methodology

4.1. *Population*

The population was comprised of undergraduate university students (year 1–year 3) and high school students (Mathayom 4–6) in the Chiang Mai province of Thailand, which is equivalent to grades 10–12 in the UK education system. The sample of students was selected and divided into two groups by using their age ranges. Consequently, the groups comprise young adolescents (Mathayom 4–6) and adult adolescents (university year 1–year 3).

4.2. *Sample*

The samples were selected by simple random sampling method (accidental sampling), with an initial sample of 125 undergraduate university students and 128 of high school students in the Chiang Mai province of Thailand. The age of majority was 19 years. After screening, valid responses were from 253 students as shown in Figure 2. The data was gathered during July to September, 2016.

4.3. *Interview*

The interviews were conducted, and the questionnaires were collected from respondents during July to September, 2016. The questions in part 1 and part 2 of Table 3 were used to evaluate perceived knowledge on cyberbullying among the

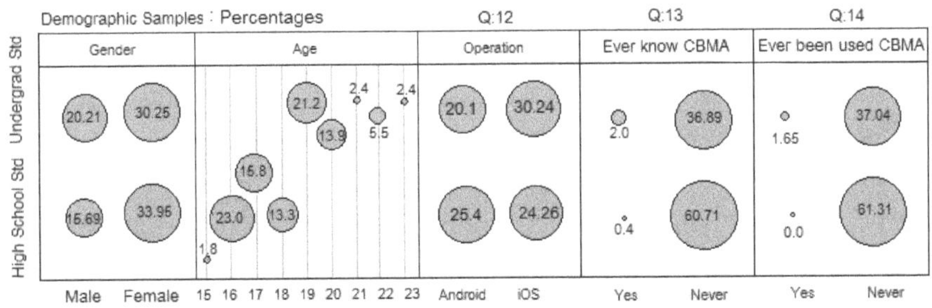

Figure 2. Demographic samples.

Table 3. List of questions for interview responders.

No.	Questions
Part 1: Accessibility	
Q1:1	Do you have a smartphone?
Q1:2	Which operation system does your smartphone have?
Q1:3	Have you ever heard about a cyberbullying application?
Q1:4	Have you ever used a cyberbullying application?
Part 2: Perception on Cyberbullying	
Q2:1	Do you know what cyberbullying is?
Q2:2	Do you know which behavior involves cyberbullying?
Q2:3	Did you used to be a victim via a social media site?
Q2:4	Did you used to be victimized via a social media site by your relative?
Q2:5	When you used to be a victim, did you your consult your family?
Q2:6	Have you ever been a cyber bully by intent or unintended purposes?
Q2:7	Do you know the punishment for cyberbullying?
Q2:8	Do you know Thailand's computer laws?
Q2:9	Do you have a tool to protect yourself from cyberbullying?

(*Continued*)

Table 3. (*Continued*)

No.	Questions
Part 3: Indicate level of attitude of the survey statements about using CBMA as a tool for education on cyberbullying	
Q3:1	CBMA can be an instrument for providing information to adolescents for consulting and educate them which decreased rate of cyberbullying
Q3:2	CBMA can provide useful information on the cyberbullying as same as professional advice from expert
Q3:3	Using CBMA can be a useful source for cyberbullying suggestions which provides protection and can stop cyberbullying
Q3:4	Using CBMA can let adolescents understand cyberbullying behavior which protects them
Q3:5	Using CBMA will not be seriously harmful to the users' adolescence
Q3:6	Using CBMA can reduce the number of victims of cyberbullying
Q3:7	Using CBMA can facilitate the users for the initial anti-cyberbullying tool
Part 4: Level of agreement about the graphical user interface (GUI) of CBMA	
Q4:1	Buttons on the screen are easy to use
Q4:2	The screen is clearly distinctive between foreground and background
Q4:3	Screen colors are appropriate
Q4:4	Each of the menu images represents understandable functionality
Q4:5	The number of menus is appropriate to use
Q4:6	Menu list is easy to use
Q4:7	Text sizes are appropriate
Q4:8	Contrast of the display is beautifully balanced
Q4:9	Users like the overall user interface design, the screen

(*Continued*)

Table 3. (*Continued*)

No.	Questions
Part 5: Level of agreement about the application usage (Experiment) of CBMA	
Q5:1	The application is easy to use and intuitive
Q5:2	The application provides useful information on the use of cyberbullying
Q5:3	Contact function is useful for the treatment and can easily access the hotline by convenience
Q5:4	Cyber photo is a useful function to inform about taking a photo to be evidence and easy to access for security
Q5:5	Sharing function can help users to share understanding of cyberbullying as well
Q5:6	The application runs quickly and stably on a mobile device platform
Q5:7	The functions of CBMA are useful and provide many benefits to adolescents but more versions need to be developed
Part 6: Level of agreement about the reliance on the use of application functionality	
Q6:1	You have confidence in the applications' instructions regarding cyberbullying
Q6:2	You believe that the application provides accurate information about cyberbullying
Q6:3	The information provided on the application helps users to understand how to protect themselves from cyberbullying and are able to use that information, similar to suggestions by experts
Q6:4	You believe that sharing cyberbullying information via social networks can protect from or discourage cyberbullying
Q6:5	You believe that CBMA will decrease the number of victims on social network sites

(*Continued*)

Table 3. (*Continued*)

No.	Questions
Q6:6	You believe CBMA can warn adolescents about cyberbullying
Q6:7	Using the application can educate you on cyberbullying
Q6:8	You accept the application and would recommend it to friends

respondents. After the CBMA was developed, the questionnaires in parts 3–5 of Table 3 were used to indicate levels of agreement of the survey statements regarding use of the CBMA. This section lets the responders try the CBMA by providing installed devices (10–15 minutes), then users were allowed to indicate levels of agreement of the following survey statements and to elicit opinions of each criterion on the graphical user interface, for example, graphics, buttons, color, size of the font, background, foreground, etc. The last section, part 6 in Table 3, includes eliciting users' opinions associated with users' confidence to use CBMA as a tool to educate themselves to understand cyberbullying.

4.4. *Questionnaire*

To determine the user satisfaction level, the questionnaire adopted a Likert scale evaluation; 5 — strongly agree, 4 — agree, 3 — neutral, 2 — disagree, and 1 — strongly disagree. This experiment assumed that ratings of 5, 4, and 3 indicated agreement and were grouped accordingly. The questions in each section follow Table 3 (excluding section 1 which is the general information of the responder). Parts 1–2 of the questionnaire aims to investigate perceived knowledge on cyberbullying, while part 3 aims to study their acceptance of using the CBMA as a tool for educating themselves. Parts 4–6 evaluate the satisfaction, performance, and reliability on the CBMA according to both groups of respondents.

5. Results

Firstly, the experiment aimed to investigate Thai adolescents about their perceived knowledge on cyberbullying. The result of the survey is shown in Figures 2 and 3 which indicate respondents' characteristics, their mobile devices, and their perceived knowledge on cyberbullying. The result shows that the majority of the respondents in this study were females, in high school, between 20 and 26. A similar percentage of high school students used both Android and iOS, whereas the majority of undergraduate students preferred iOS rather than Android. The explicit result: the majority of respondents from both groups were unaware of the cyberbullying application. To demonstrate the level of perceived knowledge of adolescents, Figure 3 shows the percentage of adolescents in bubble plot matrices for the questions corresponding to the list in Table 3, part 2. The percentages of perception were compared between both groups.

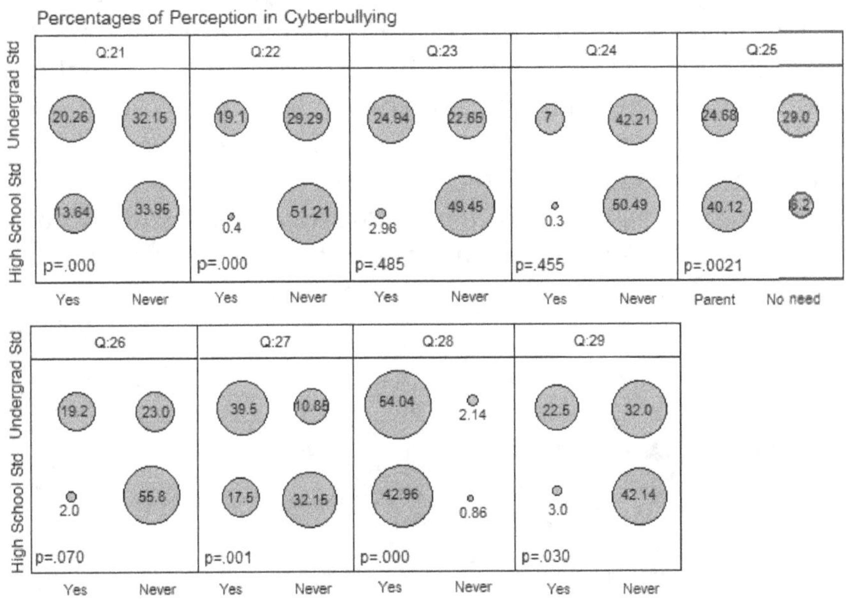

Figure 3. Bubble plot matrices show the results of questions listed in Table 3, part 2.

The results show that the knowledge perception between both groups were different in each question. The undergraduate student group had a higher percentage of perception than the high school student group for most questions (Q2:1, Q2:2, Q2:5, Q2:7, Q2:8, and Q2:9), except for questions Q2:3, Q2:4, and Q2:6. This means both sample groups did not understand cyberbullying. Interestingly, the results show that both groups agree to consult their families when they encounter cyberbullying. The point that needs attention is in Q2:5; it shows about 29% of undergraduate university students and 6.2% of high school students, respectively, will not ask their families. When considering the proportion, the result of this question needs to be considered.

The next result answers objectives 2 and 3. It shows the bubble plot matrices of the average of attitude toward applying the cyberbullying application as a tool for adolescents (corresponding to the questions in Table 3, part 3). The result indicates users' satisfactions of the CBMA after a trial which is shown in three parts, as following: the first one is the attitude toward the use of cyberbullying application as a tool for adolescents to gain knowledge and consultant on cyberbullying. The result shown in Figure 4 indicates the average point corresponding to the satisfaction levels using the Likert scale. The averages of users' attitudes for the high school students are higher than the undergraduate students in all criteria (exclude Q3:2 and Q3:5 which are not different).

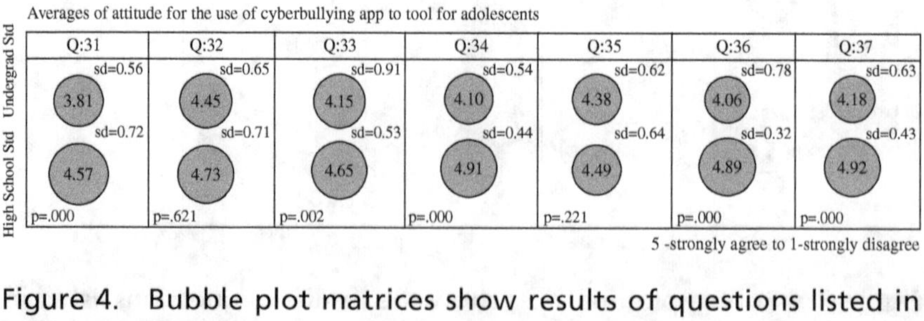

Figure 4. Bubble plot matrices show results of questions listed in Table 3, part 3.

The second result shown in the bubble plot matrix shows the averages of user satisfaction with the CBMA about the GUI (questions in Table 3, part 4). The average of user satisfaction on the GUI has been collected after users have tested the CBMA, which is shown in Figure 5. From the result of Q4:1, Q4:3, Q4:4, Q4:5, Q4:6, and Q4:8, the corresponding buttons, screen colors, menu images, number of menus, menu list, and screen contrast have to be improved, because the CBMA is expected to be an appropriate tool for both groups.

The third result: the bubble plot matrices show the averages of user satisfaction of the CBMA system (questions in Table 3, part 5). Figure 6 indicates the averages of user satisfaction over the CBMA system. The questions in this section aim to investigate users' understanding for each function in the CBMA. The results show both groups appreciated the CBMA functionalities with different satisfaction levels between two groups. The interface of the application was agreed to be user-friendly by both groups, which is indicated in question Q5:1.

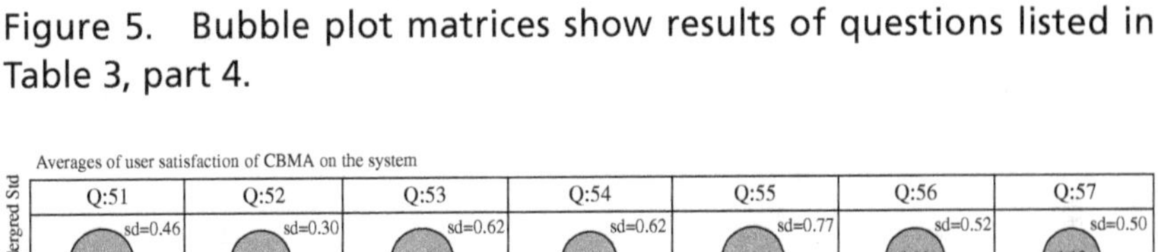

Figure 5. Bubble plot matrices show results of questions listed in Table 3, part 4.

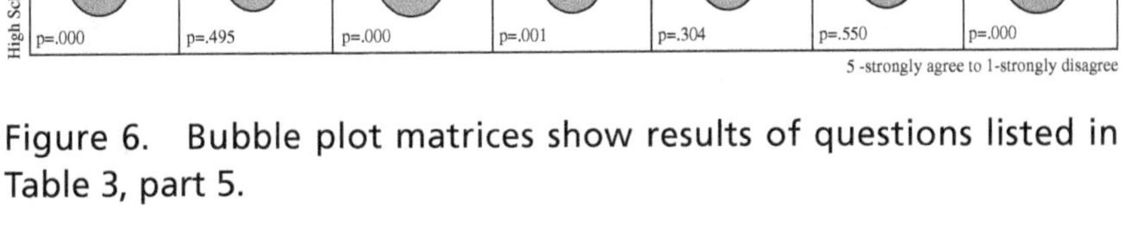

Figure 6. Bubble plot matrices show results of questions listed in Table 3, part 5.

There is an interesting point that shows a high satisfaction level in Q5:3. It shows high school students appreciated the function that provides direct calls for consultation via phone. Whereas, the function of the cyberbullying infographic sharing was much appreciated by undergraduate university students (Q5:5). The result in Q5:6 indicates a higher satisfaction level for the undergraduate university students than the high school students.

However, the result for the question Q5:1, which is "application is easy to use and intuitive" shows that the CBMA still needs to be improved because the satisfaction levels between two groups were different. Similarly, the results of Q5:3 (contact function), Q5:4 (cyber photo), and Q5:7 (functions of the CBMA) show that user satisfaction between two groups were different, but the high school student group has a higher satisfaction than the undergraduate university student group.

In the next section, the bubble plot matrices show averages of user confidence in the use of the CBMA (questions in Table 3, part 6). Figure 7 shows the level of user confidence when using the CBMA to be a tool for educating and encouraging adolescents to understand cyberbullying. The results show the group of high school students has a higher confidence level than the group of undergraduate university students, except for questions Q6:1, Q6:3, Q6:4, Q6:6, Q6:7, and Q6:8 which were not different between both groups. The highly different confidence level was found in Q6:2 and Q6:5, which indicates that high school students had higher levels of confidence in the use of

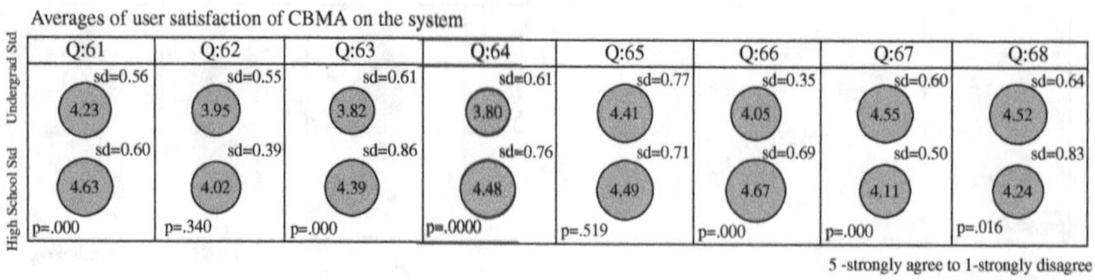

Figure 7. Bubble plot matrices show results of questions listed in Table 3, part 6.

information from the CBMA than undergraduate university students and in the suggestions by experts, as well as the belief in the use of the CBMA will be able to decrease the number of victims on SNS.

Moreover, the confidence on sharing cyberbullying information via the social network can discourage cyberbullying behavior. The high school students have greater confidence than undergraduate university students and that is shown in Q6:4, whereas the belief in the use of the CBMA can decrease the number of victims on SNS, and the result shows the group of high school students has greater confidence than the undergraduate university students.

However, regarding the results in Q6:7 and Q6:8, the confidence of undergraduate university students was higher than that of the high school students when we asked them about using the application to educate themselves to understand cyberbullying, as well as the acceptance of the application, and would they recommend it to their friends, respectively.

Figure 8 shows the overall comparison results of testing differentiation between two groups (averages from parts 3–6 in Table 3). When evaluating the overall attitude of using the CBMA as a tool to educate and encourage adolescents on cyberbullying (part 3), the results between two groups were different at a statistical significance of 0.01. The results also show *p*-value <0.05, which rejects H_0 and accepts H_1. Therefore, the overall attitude of using the CBMA between two groups

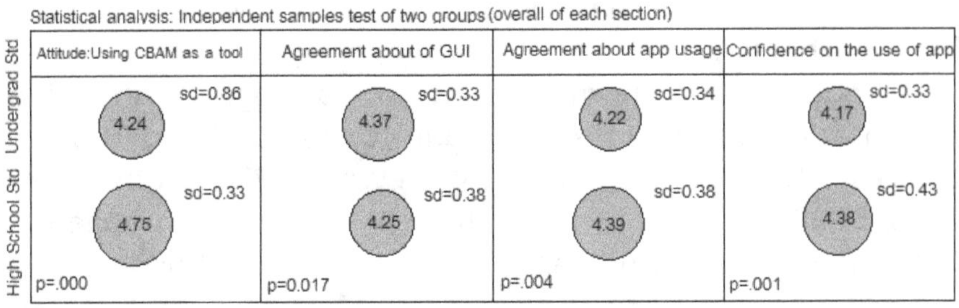

Figure 8. Comparison differentiation overall in each section (parts 3–6).

were different, where the high school students had a higher level of attitude than the undergraduate university students.

Regarding the overall agreement concerning the GUI (part 4), the results of the two groups were different with *p*-value <0.05, which rejects H_0 and accepts H_1. The undergraduate university students had a higher level of attitude than high school students. Regarding the agreement about the application usage and the confidence when using the application (parts 5 and 6), the results of the two groups were different. The results also indicate *p*-value <0.05, which rejects H_0 and also accepted H_1, where the high school students had a higher level of agreement and attitude than the undergraduate university students.

6. Discussion

To achieve the first objective of the research for investigating Thai adolescents' perceived knowledge on cyberbullying. The result in the questionnaire (part 1) indicates the demographic of samples (the first-time survey) which highlights their characteristics. The majority of sampled students had never used any CBMA and had never known about the existing tools for anti-cyberbullying. The result in part 2 of Figure 3 shows the evidence from the first round of the survey which found that the concerning issue was lack of knowledge of cyberbullying. Corresponding to the studies of Tapanya [2006], Laeheem *et al.* [2008], and Sittichai and Smith [2015], the result of this survey informs that Thai adolescents lack knowledge regarding cyberbullying.

Although there are many tools for learning and providing consultation on or protection against cyberbullying in many countries [Menesini *et al.*, 2015; Brewer, 2011; Jacobs *et al.*, 2014; STOPit cyberbully, 2015], they have not been applied in Thailand yet, due to the language barrier. So, in this research, we need to develop an application for Thai adolescent usage by solving the language barrier problem and providing suitable functions for learning and consultation or guidance on cyberbullying.

The second research objective is to develop the CBMA which is based on iOS operating system. The CBMA was developed under the conceptual framework for being a tool to provide guidance when encountering a cyberbullying problem which followed suitable suggestions of many researchers. These included the useful suggestions of Royen *et al.* [2015], Nadia *et al.* [2015], GroupCast LLC [2015], Menesini *et al.* [2015], Brewer [2011], and Jacobs *et al.* [2014], which mentioned that the system should have a function to help victims, should provide suitable information resources, and should have a consultancy function. The complete version of the CBMA was released for public use after useful functions were applied including the function for providing cyberbullying contents and the function for calling to consultants. These functions may have similar purposes to applications like No-Trap, Web Quest, and Online "Pestkoppenstoppen" [Menesini *et al.*, 2015; Brewer, 2011; Jacobs *et al.*, 2014], because these functions also aim to help adolescents to answer cyberbullying questions.

The third research objective is to evaluate user feedback regarding their attitude, GUI satisfaction, system performance, and user confidence on the CBMA. About the user attitude, the result of the questionnaire is shown in Figure 4, which was collected from respondents (in the second-time survey of the same samples) after a trial of the CBMA as a tool for anti-cyberbullying. The result claims that the users' attitude towards the use of the CBMA as a tool for educating on cyberbullying was satisfactory for both groups. Although, the result of testing the hypothesis between two groups were different, the study results show that the CBMA could be more appropriate for the high school students than undergraduate students.

However, the result shows good attitude among users if they were using for primary help when encountering a cyberbullying problem. It was no more than expected and the attitude of users was appreciated on the CBMA because, in Thailand, no application or any suitable tool for anti-cyberbullying was available previously. There are only websites or TV shows in Thai government channels which appeal to Thai adolescents regarding the issue. So, expectations of the new tool may have effect on their attitude toward research.

In terms of user feedback, the result shows that the majority of users are satisfied with the CBMA regarding its GUI. The menu buttons, screen size, and colors were designed appropriately. Due to mobile devices being popular among Thai adolescents and cyberbullying content also developed on the mobile platform, the new cyberbullying content merged with a suitable GUI, which encouraged the respondents' satisfaction on the CBMA GUI as shown in the result. The results of testing the hypothesis between two groups were different, and the study points out the suggestion of what undergraduate students appreciated, which is useful for the development of the next version of the CBMA.

The application usage result shows each of the functionalities is useful for adolescents, for example, function for providing the cyberbullying content, function of contacts, cyber photos sharing function, etc. Some of these functions are similar to the well-known applications such as STOPit and GroupCast [STOPit cyberbully, 2015; GroupCast LLC, 2015]. However, since the result of testing the hypothesis between two groups were different, for the next version of the CBMA, the developer should find more specific functions to meet undergraduate student requirements.

The last result subjects for the third research objective, which indicates the level of agreement among users regarding the user reliance on the CBMA functionalities. When focusing on the infographic sharing function, it may probably be a good presentation of the CBMA because users are able to share or post cyberbullying infographic sharing content. The function could create awareness of cyberbullying for others very well. Although, posting cyberbullying infographic sharing content cannot be automatically monitored by the cyberbullying system as suggested by Royen *et al.* [2015], the CBMA can help promoting awareness of cyberbullying phenomenon, and indicate cyberbullying issues and share them on the Facebook wall, with others.

In comparing both groups, the group of high school students appreciated and had a higher agreement level than the group of undergraduate students when trying the application. This shows that the application may probably be an appropriate tool for adolescents in high school level rather than university level, since the contents or the functions were suitable for their age range. In Thailand, there have been are several studies relating the relationship between cyber victimization and age range. The highest number of offenses occurs between 13 and 15 years old (high school level) [Tokunaga, 2010; Tapanya, 2006]. If the proposed application is released, the problem of cyberbullying among Thai adolescents could be reduced or, at least, learning capability on cyberbullying knowledge would be unrestricted to the language barrier. As a result, this is a good opportunity for promoting the CBMA as a tool to encourage anti-cyberbullying among Thai adolescents.

Finally, the application may boost the conceptual ideas of the study of the following researchers [Nadia *et al.*, 2015; Menesini *et al.*, 2015; Brewer, 2011; Jacobs *et al.*, 2014] to spread wider in Thailand than before.

7. Conclusion

This research aimed to investigate perceived knowledge on cyberbullying among adolescents, to develop the CBMA based on iOS operating system and to evaluate users' feedback regarding GUI, system performance, confidence, and attitude.

The results show that Thai adolescents lack perceived knowledge regarding cyberbullying. Moreover, it also shows that high school students have lower perceived knowledge than undergraduate students. The majority of respondents were lack of knowledge on the cyberbullying.

The CBMA has been fully developed and released in the Apple Store. After the trial of the CBMA, a survey regarding GUI, system performance, users' confidence, and attitude on the CBMA was conducted. The result of testing the hypothesis of user feedback regarding the GUI of the CBMA, system performance, confidence, and attitude show that users from both groups were different. The CBMA may not be appropriate enough for the undergraduate students, but it could be a useful tool for the high school students. The explicit outcome indicates the high school students agreed on the use of the CBMA as a tool to educate themselves on cyberbullying and to be aware of cyberbullying instances. Although, the CBMA is more appropriate for high school students than undergraduate students, and the challenging features and functions to suit Thai adolescents should be considered for the next version development of the CBMA.

This is the first cyberbullying application for Thai adolescents developed in Thai language, and mobile application is proven as a suitable channel for providing information on cyberbullying for Thai adolescents, especially a high school student. It is an excellent opportunity to facilitate cyberbullying literacy to Thai adolescents using the CBMA.

Acknowledgments

This research was supported by the Embedded Systems and Mobile Application Laboratory. The authors would like to thank the College of Arts, Media and Technology, Chiang Mai University for the funding of this research.

References

Agatston, P. W., Kowalaski, R. and Limber, S. (2007). Students' Perspectives on Cyber Bullying. *Journal of Adolescent Health*, **41**: 59–60.

Bianchi, A. and Phillips, J. G. (2005). Psychological predictors of problem mobile phone use. *CyberPsychology and Behavior*, **8**(1): 39–51.

Brewer, E. A. (2011). Fighting fire with fire: The use of a multimedia WebQuest in increasing middle school students' understandings of cyberbullying. Thesis and Dissertations, Loyola Marymount University School of Education Los Angeles, CA, USA.

Cassidy, W., Faucher, C. and Margaret, J. (2013). Cyberbullying among youth: A comprehensive review of current international research and its implications and application to policy and practice. *SAGE Journals*, **34**:575–612.

Charles, E. N., Sharon, P. and Jessica, R. (2013). Cyberbullying: A review of the literature. *Universal Journal of Educational Research*, **1**(1):1–9.

Ehrenberg, A., Juckes, S., White, K. M. and Walsh, S. P. (2008). Personality and self esteem as predictors of young people's technology use. *CyberPsychology and Behavior*, **11**(6):739–741.

Elizabeth, W. and Robin, M. K. (2015). Cyberbullying via social media. *Journal of School Violence*, **14**:11–29.

GroupCast LLC (2015). CyberBully hotline. Available at http://www.cyberbullyhotline.com/mobile-app.html.

Hinduja, S. and Patchin, J. W. (2010). Cyber-bullying research summary — Cyber-bullying and suicide. Cyber-bullying Research Center. Available at http://cyberbullying.org/cyberbullying_and_suicide_research_fact_sheet.pdf.

Huang, Y. Y. and Chou, C. (2010). An analysis of multiple factors of cyberbullying among junior high school students in Taiwan. *Computers in Human Behavior*, **26**:1581–1590.

Jacobs, N. C. L., Volloink, T., Dehue, F. and Lechner, L. (2014). Online Pestkoppenstoppen: Systematic and theory-based development of web-based tailored intervention for adolescents' cyber bully victims to combat and prevent cyberbullying. *BMC Public Health*, **14**:396.

Kowalski, M. R., Morgan, C. A., Drake-Lavelle, K. and Allison, B. (2016). Cyberbullying among college students with disabilities. *Computer in Human Behavior*, **57**:416–427.

Kwan, G. C. E. and Skoric, M. M. (2013). Facebook bullying: An extension of battles in school. *Computers in Human Behavior*, **29**:16–25.

Laeheem, K., Kuning, M., McNeil, N. and Besag, V. E. (2008). Bullying in Pattani primary schools in southern Thailand. *Child: Care, Health and Development*, **35**:178–183.

Lenhart, A., Madden, M., Smith, A., Purcell, K., Zickuhr, K. and Rainie, L. (2011). Teens, kindness and cruelty on social network sites. How American teens navigate the new world of "Digital Citizenship". *Pew Research Center's Internet & American Life Project*. Washington, DC.

Li, Q. (2008). A cross-cultural comparison of adolescent' experience related to cyberbullying. *Educational Research*, **50**(3):223–224.

Menesini, E., Palladino, B. and Nocentini, E. (2015). Online and school based intervention to prevent cyberbullying among adolescents. *Cyberbullying from Theory to Intervention*. Taylor & Francis, Switzerland.

Mishna, F., Khoury-Kassabri, M., Gadall, T. and Daciuk, J. (2012). Risk Factors for involvement in cyberbullying: Victims, bullies and bully–victims. *Children and Youth Service Review*, **34**:63–70.

Nadia, A. W., Wan, A. J. W. Y. and Balakrishman, M. (2015). The use of multimedia in increasing perceived knowledge and awareness of cyber-bullying among adolescents: A pilot study. *Procedia Social and Behavioral Sciences*, **176**:745–749.

National centre Against Bullying (2016). Types of bullying. Available at https://www.ncab.org.au/bullying-advice/bullying-for-parents/types-of-bullying.

Noam, L. L. and Michal, D. C. (2015). Comparing cyberbullying and school bullying among school students: Prevalence, gender, and grade level differences. *Social Psychology of Education*, **18**(1): 1–16.

Online Apple Store (2016). Cyberbullying: Application in apple store. Available at iOS.

Patchin, J. W. and Hinduja, S. (2006). Bullies move beyond the schoolyard. A preliminary LOOK at cyberbullying. *Youth Violence Journal*, **4**(2):148–169.

Royen, K. V., Poels, K., Daelemans, W. and Vandebosch, H. (2015). Automatic monitoring of cyberbullying on social networking sites: From technological feasibility to desirability. *Telematics and Informatics*, **32**(1):89–97.

Smith, P. K., Mahdavi, J., Carvalho, M., Fisher, S., Russell, S. and Tippet, N. (2008). Cyberbullying: Its nature and impact in secondary school pupils. *Journal of Child Psychology and Psychiatry*, **49**(4):376–385.

Sittichai, R. and Smith, P. (2015). Bullying in South-East Asian countries: A review. *Aggression and Violent Behavior*, **23**:1–15.

STOPit cyberbully (2015). Available at http://stopitcyberbully.com.

Tapanya, S. (2006). *A survey of Bullying among Students*, Psychiatric Department, Faculty of Medicine, Chiangmai University (in Thai), Chiangmai.

Tokunaga, R. S. (2010). Following you home from school: A critical review and synthesis of research on cyberbullying victimization. *Computers in Human Behavior*, **26**:277–287.

Willard, N. (2005). Educators guide to cyberbullying and cyber threats. Available at http://new.csriu.org/cyberbully/docs/cbteducator.pdf.

Wolak, J., Mitchell, K. J. and Finkelhor, D. (2006). Online victimization of youth: Five years later. National Center for Missing & Exploited Children. Available at http://www.unh.edu/ccrc/pdf/CV138.pdf.

www.ingramcontent.com/pod-product-compliance
Lightning Source LLC
LaVergne TN
LVHW010330220926
848484LV00004BB/870